D0260904

WILD FLOWERS IN DANGER

By the same author

THE ORIGINS OF GARDEN PLANTS

MR. MARSHAL'S FLOWER ALBUM

THE COMPANION TO ROSES

WILD FLOWERS IN DANGER

John Fisher

with photographs by the author

LONDON
VICTOR GOLLANCZ LTD
1987

To my wife
for giving me licence to explore

First published in Great Britain 1987
by Victor Gollancz Ltd,
14 Henrietta Street, London WC2E 8QJ

British Library Cataloguing in Publication Data
Fisher, John, *1909–*
Wild flowers in danger.
1. Rare plants
I. Title
582.13 QK86.A1

ISBN 0–575–03893–4

Typeset at The Spartan Press Ltd, Lymington, Hants
and printed in Great Britain by
St Edmundsbury Press Ltd, Bury St Edmunds, Suffolk
Colour illustrations originated and printed by
Acolortone Ltd, Ipswich

Contents

Illustrations

Introduction

Few plants, when examined closely, are completely without beauty, but those at risk have a special attraction for the botanist, who is conscious of an irresistible longing to see them, if only just once, before it is too late. But he is inspired at the same time by a desire to protect them, to bequeath to his descendants as much as possible of the living world which he himself has inherited. He is possessed of the sentiment that leads a man to plant an acorn knowing that he may never see the magnificent oak which he hopes will grow from it.

These two impulses, to admire and to protect, are inseparable, and it is to be hoped that neither the one nor the other predominates in what follows.

But, on balance, should anyone write anything at all about rare plants? Many botanists sincerely believe that to do so is to encourage fresh battalions, some equipped perhaps with garden trowels, to go on safari. They see publicity as an added threat to species already at risk, and to their habitats. But, equally, such plants cannot be protected adequately without the support of hundreds, perhaps thousands of volunteers who subscribe to Conservation Trusts, raise funds, organise meetings, and keep records for posterity. Their interest must be aroused and their enthusiasm stimulated by knowledge of the botanical treasures they are to help protect. They must be encouraged to look for and record not only rare plants on the sites that are already known but also rarities on new sites as yet undiscovered; and they may often be the first to sound a warning about an unforeseen danger posed to a vulnerable site by some new development.

Unless these enthusiasts can recognise these rare plants in their natural surroundings, and are aware of the kind of habitats they frequent, they can hardly be expected to realise what is before their eyes, or to possess the courage and authority to tell their local Conservation Trust when they have found a rarity.

Despite all the development and building that has disfigured so much of Britain there are areas in almost every county that are relatively unexplored by those most qualified to do so, namely the people living in the immediate neighbourhood. They will be the first to chance on the rare casual, the unusual hybrid, the well-known species in an unexpected locality, or perhaps even a species new to Britain.

It is for them that this book has been written.

ACKNOWLEDGEMENTS

No botanist is entirely self-sufficient. He depends on friends who have been where he has not been and have seen what he has not seen, and, though here and there, he may be able to make a contribution to the common weal, he is greatly outnumbered by others offering to help him. In my case some of the most generous have been companions in the field, frequently in adversity; others I have met only on the telephone, through the post, or at conferences. A few, though no longer alive, I prefer not to describe as 'the late . . .' I am permanently indebted to them all, and if any others have been omitted it is surely through inadvertence, not ingratitude. And so, in alphabetical order they are: Tony Aldridge, Doris Ashby, Ron Boniface, Lady Anne Brewis, Mary Briggs, Margaret Burnhill, Breda and Ernie Burt, Ray Clarke, Dr P. Cribb, John and Chris Dony, Joan Duncan, Kathy Fallowfield, Lynne Farrell, Ian Finlay, George Forster, Dr L.C. Frost, Mrs J.L. Godfrey, David Grant, Charlie and Thelma Hall, Arthur G. Hoare, Sonia Holland, Florence Houseman, David Mardon, Mary McCallum Webster, Edgar Milne-Redhead, Harriet Muir, J. Partridge, Vera Paul, John Port, Margaret Reese, Dr Reynolds, Elizabeth Rich, Tim Rich, Marion Riddell, Mark Rutterford, P.J. Shaw, Bill Shepard, Joyce Smith, Derek Steeden, Mrs A.J. Stevens, Joyce Stewart, Joan Swanborough, Leo Taylor, Janette Thorne, Dr G. Tranter, Tim Westlake, John Workman and Barry Yates. My warm thanks go to Anne Payne for her immaculate computerised typing.

J.F.

THE BACKGROUND

— 1 —

Some Missing Plants

To see how many plants have vanished over the centuries we need only to examine two records of the county of Kent; one prepared in 1629 and the other record for the same county published in 1982, a spread of little more than 350 years.

The earlier record of wild flowers in the 'Garden of England' was kept by Thomas Johnson, the apothecary, who was born probably at the beginning of the seventeenth century and died in 1644 from a wound in the shoulder which he sustained while fighting for King Charles I against Cromwell's men.

Already, during a journey in the northern counties of Britain, Johnson had discovered and described plants that had not been previously recorded including *Anagallis tenella*, Bog Pimpernel, *Stratiotes aloides*, Water Soldier, and *Gentiana pneumonanthe*, Marsh Gentian.

Then, in 1629, he undertook Iter Plantarum Investigationis – a journey for the investigation of plants in the county of Kent with nine other Fellows of the Society of Apothecaries. It was, he wrote, a praiseworthy custom that had grown up a few years earlier among students of herbalism to undertake twice a year or more a journey often lasting three or four days with the object of discovering new plants. In 1629, the Society decided to investigate part of the Kentish countryside and so, early on the morning of 13th July, the appointed day, the ten good companions met in the churchyard of St Paul's Cathedral and set off. Johnson's account of his journey was written in the language of scholars: Latin, but an excellent translation has recently been published (see Bibliography on p. 185.) The party set off not by road, but instead hurried down to the Thames and hired two boats to take them downstream with the tide. The weather, however, worsened considerably and one boat with four of the party had to put into Greenwich and tarried there to refresh themselves. The more determined members of the party proceeded directly to Gravesend and, after having eaten at an inn, left a note for the laggards, and moved on along the highway to Rochester, where it was decided to spend the night. On the walk there, of roughly ten miles, they noted Thorow-wax (*Bupleurum rotundifolium*), White Horehound (*Marrubium vulgare*), Corncockle (*Agrostemma githago*), Cornflower (*Centaurea cyanus*), Field Gromwell (*Lithospermum arvense*), Shepherd's Needle (*Scandix pecten-veneris*), Catmint (*Nepeta cataria*) – among 104 others. The follow-up party, who had landed at Erith and walked from there to Gravesend, found in addition Wild Liquorice (*Astralagus glycyphyllos*).

From Rochester they walked up over the hill to Gillingham where in the churchyard they saw *Sambucus ebulus*, Dwarf Elder, growing in profusion. En route for Sheppey they found *Dianthus armeria*, the Deptford Pink.

Johnson and one other of the party, Jonas Styles, set off westwards across country from Rochester to Cliffe where they spent the night. En route they collected nearly 40 more plants. In the Dartford area they came across what appears to have been *Verbascum lychnitis*, White Mullein. Some of them came back aboard a ship returning from the East Indies and were rewarded with a big Indian nut and piece of sugar cane.

To see what had happened during the three and a half centuries that have passed since that walk we must consult *The Atlas of the Kent Flora* by Eric G. Philp, F.L.S. (see Bibliography). The *Atlas* includes the results of a survey lasting ten years from the 1st January 1971 to 31st December 1980. For this purpose the county was divided into 1,044 areas measuring 4 square kilometres of which Johnson's groups covered perhaps 60.

As was to be expected, the rarities have disappeared. Thus Thorow-wax, *Bupleurum rotundifolium*, was not recorded anywhere in Kent during the survey for the Atlas. *Agrostemma githago*, Corncockle, had vanished: *Centaurea cyanus*, Cornflower, was encountered only as an escape from cultivation: as was *Marrubium vulgare*, White Horehound. *Lithospermum arvense*, Field Gromwell, had vanished from the county. *Dianthus armeria*, the Deptford Pink, had vanished from where Johnson had found it.

What is more surprising, however, is the disappearance from the area covered by the Johnson expedition of some of the more common chalk-loving plants which the apothecaries had found, such as *Campanula rotundifolia*, Harebell; *Campanula glomerata*, Clustered Bellflower;

Misopates orontium, Lesser Snapdragon; and *Galeopsis angustifolia*, Red Hempnettle. A sad story.

Since those days strenuous efforts have been made to discover which plants are most at risk, and we shall first ponder the question how the degree of risk to a given species is now assessed.

— 2 —

Wild Flowers – How Rare is 'Rare'?

Flowers at risk: which ones should be protected and which don't really qualify?

Conservationists – who would like to save all plants except a few hyper-invasive ones such as Japanese Knotweed – realise that they are wise to restrict their efforts to plants that are truly at risk and that the public and politicians are more ready to pay heed to their appeals provided that their list of threatened species is not unmanageably long.

Rare plants are clearly at greater risk than common ones. Thus if the population of a species is restricted to a single site, it is more likely to be wiped out by a single disaster – a flood, a fire or a plough – than a species which has many different sites as back-up. Also, a widely distributed species has a better chance of survival because if there are only a few of a kind, the chances of improving the breed will be correspondingly reduced. Again, if a species has survived in but a few localities, the reason is probably that it has been unable to withstand competition elsewhere.

Thus the rare flowers must be regarded as being at high risk. But where should the line be drawn?

Rarity according to the *Oxford English Dictionary* can mean 'few in number and widely separated from each other (in space or time); forming a small and scattered class'. But how 'widely' is 'widely separated'?

When we look into matters more closely we begin to see that rarity is a matter of how much space we are considering. *Helianthemum apenninum*, White Rock Rose, for example, may be abundant on Brean Down, but it is, nevertheless, a rare plant, though it is also to be found in Belgium, France, West Germany, Spain, Portugal, North and Central Italy, Albania, Greece, Crete, Asia Minor and Morocco. In England the Spring Gentian is a relatively common plant in Upper Teesdale but nowhere else in Great Britain. In continental Europe however, the

Gentian grows on the mountains of central Europe southwards to the Sierra Nevada in Spain, in Italy on the Apennines, in Albania, and the Caucasus, and sub-species are found in north and central Asia and Morocco. So on a wider scale neither species is really so rare as to be endangered.

The International Union for Conservation of Nature and Natural Resources (IUCN) recognises four main categories of risk.

Plants most at risk are described as 'Endangered'. These include for example plants threatened with extinction because their numbers have been reduced to a critically low level, casting doubts on their ability to continue breeding successfully, or because the plants or their environment are in imminent danger from man's activities.

Plants described as 'Vulnerable' are those likely soon to become 'Endangered' if measures are not taken to protect them. Their population may be tending to decrease, or their habitat – if not already threatened – is likely to become so, or they are sufficiently scarce for it to be advisable to take measures to protect them.

Plants described as 'Rare' are not under immediate threat but are thinly scattered and limited in their habitats and, while they may not require immediate protection, deserve to be monitored, so that any sharp decline in their numbers or danger to their habitat from fire, flood, or subsidence will be notified to conservationists. The fourth category is 'Extinct'.

The Pocket Guide to Wild Flowers by David McClintock and R.S. Fitter, assisted by Francis Rose, uses a star system to indicate which plants are rare or uncommon – to add to the pleasure of discovering a rarity – and to alert field botanists to the possibility that a flower that they have found growing abundantly in many different localities is unlikely to be a rarity awarded three stars. In McClintock's book no stars are awarded to a plant that is common and widely distributed. One star is awarded to a plant which is only locally common. Two stars are for scarce plants which grow regularly in limited areas (Crested Cow-wheat is one) or for plants that may be thinly scattered over a wide area; while three stars are for rarities growing in only a few places and which may be rare even there.

But when a case has to be put to Parliament for a rare plant to be given protection, the question is naturally asked: 'Can you provide figures?' One way of providing them is shown in *The British Red Data Book* compiled by F.H. Perring and L. Farrell (see Bibliography). This is an indispensable form-book for anyone seeking to know the risk of extinction for any particular plant. It is the first of a series on plants and deals with Vascular Plants, that is those equipped with a system of vessels through which liquids can be carried from one part of the plant to another. (In other words the term covers most of the ordinary plants which have a framework of root, stem and leaves, in contrast to the mosses and their close relations.)

Both authors are experts in their own fields. Franklyn Perring is the indefatigable General Secretary of the Royal Society for Nature Conservation. Lynne Farrell, a high-ranking botanical scientist, is Secretary of the Permanent Working Committee on Conservation of the Botanical Society of the British Isles, and is a specialist at the Nature Conservation Council on the ecology of rare plants.

The British Red Data Book (which has the backing of the Nature Conservancy Council and World Wildlife Fund) is about flowers that are here today – and may be gone tomorrow – as well as about a few absentees that have never re-appeared in the wild. Some of the plants are distressingly beautiful. Others, though less aristocratic, have personality – characteristics which distinguish them sharply from any other species. A few have an appealing fragility. Some, without a touch of class, have only their scarcity value to recommend them. And some we love all the more because of the knowledge that they are probably already doomed.

By comparing new 10 km square findings with older records it became possible to estimate the rate of decline and therefore one further aspect of the risks to which each species is subject. Both 10 km square records and those of individual sites correspond with grid map references and have been sent to the central Biological Records Centre at the Monks Wood Experimental Station at Huntingdon. The authors of *The British Red Data Book* also devised a more detailed system for assessing the total degree of risk faced by each rare species.

Plants found on 15 or fewer sites have been awarded a Threat Number ranging between 1 and 4. Species that are declining rapidly – 33 or more per cent – during the period covered have a Threat Number rising from 0 to 2.

Additional Threat Numbers are added to plants that might be considered attractive to gardeners (up to 2 points), and further points are added to species outside nature reserves, with extra points in cases where both the sites and the plants on them are readily accessible.

Thus, under this system a plant that is attractive to gardeners, that had been declining rapidly and is now to be found only in one or two easily accessible sites, none of which are in a Nature Reserve, with the plants themselves easily approachable (i.e. not halfway up a cliff or at the end of a 10-mile walk) could theoretically earn a total Threat Rating of 15.

So far, the most threatened plants recorded have a rating of 13 – one of them being *Epipogium aphyllum*, the Ghost Orchid, fully described here. (Some of the others are far less exciting.)

Many factors have to be borne in mind each time the list is revised. Around 100 species already in *The Red Data Book* came up for consideration during 1986. The first question to be asked is: How many species can in practice be protected by regulations enforcible at law? What is a reasonable number that can be justified objectively in Parliament? The next question is whether the species to be protected can be readily identified. Many reeds, sedges, grasses and hawkweeds cannot.

Next come the priorities. Should a plant that is attractive – and therefore more likely to be picked or trampled on – be given preference over a less attractive species which happens to be botanically important – or perhaps exclusive to the British Isles? How much protection should be given to plants that are common in the British Isles but are rare in continental Europe? Should rare plants in Britain be protected if the same species is common on the continent? How should species such as Henbane and Corn Marigold that are dependent for their survival on disturbed ground be protected?

But extra research is needed if we are to see which plants, if not already so, are likely to become rare.

Gentiana verna is one of many plants of which the life-cycle, as well as population, has been regularly recorded. But scientists rightly believe that head-counting by itself is not a sufficient basis for predicting the future of a rare species. For example, we need to know whether an increase in population is due to some old plants living longer or to more new plants coming on. The life expectancy of plants can be charted only by logging individual plants. Thus visitors to a site containing the *Saxifragia cernua*, Drooping Saxifrage, are likely to notice coloured pins close to the plants such as are normally seen on military maps. These literally pinpoint the history of each plant. In other areas where security is vital, metal markers are buried in the ground to be discovered later with the help of metal detectors. Eventually, perhaps over a ten-year period, statistics of life expectancy and mortality can be studied in the same way as life assurance actuaries assess the future of their fellow humans. When detailed studies are made of the dynamic population of a species, the selected site can be marked off into areas of about one foot square, with permanent metal boundary corners in the ground. Then, each year at the appropriate time, a recorder visits the site taking along a portable grid which matches the area between the permanent markers. The grid is numbered along each of its sides with reference numbers like those along the edge of a map, which makes it possible for the observer to record which plants and how many of them are growing in each of the separate small squares covered by the grid.

In Upper Teesdale where the Gentian – among other rarities – grows, studies in population dynamics have been carried out by a clutch of scientists based on Durham University.

Along with life-expectancy tables, studies are made to determine what influences life expectancy. Thus rabbit-grazing reduces the population of gentians, especially after a mild winter (when the rabbit population increases) and also during a dry summer when almost anything green will be eaten. On the other hand, certain rare plants – such as *Iberis amara*, Wild Candytuft, prefer to grow in areas scraped clear of other vegetation by the rabbits, and orchids on the downs grow best when rabbits eat down the coarser grasses. Again, grazing by cattle and sheep may be beneficial if this prevents scrub from monopolising grassland.

Annual plants are more difficult to monitor than perennial. The population varies more sharply from one year to another than is the case with perennials. And sometimes the seeds lie buried for years in the soil seed-bank before making an appearance, as happens in particular with *Chrysanthemum segetum*, Corn Marigold.

Ideally in this kind of investigation the whole of the life-cycle will be researched. The percentage of plants that flower, and those that produce the seed . . . the numbers of seeds, the method by which they are dispersed . . . the numbers to be found in the soil seed bank, how long the seeds remain viable, the numbers that germinate, and finally the proportion that succeed in establishing themselves in variable conditions of temperature and rainfall. But even within these limits there can be no hard and fast rule. Periodically some plants may qualify for removal from the list because plants subsequently have been discovered in areas where they had not been recorded previously. Or new plants are discovered that were previously unrecorded in Britain and must therefore be added. Moreover there could be more than one opinion about many different species. A plant which some botanists might regard as attractive and therefore more at risk might be considered by other botanists to be totally without interest. And so on.

To compile a list of 100 species for this book from a list of 317 has not been easy. But some are more easily eliminated than others. For instance rare sedges, grasses and reeds, while of concern to advanced botanists, may not interest the average first timer and seldom provide striking photographs. Some other plants are to be seen in their native state only on remote mountains or bogs to which few readers will want to make a separate journey.

Then there are a few other plants such as Green Hellebore which though not of extreme rarity are of such beauty (in my opinion) that they are in just as much danger as some of the less attractive and rarer species. And their disappearance would certainly be even more of a loss. So they have been included. And, if a few are still missing, this shortfall must be ascribed to the author's own personal prejudices.

— 3 —

The Sanctuaries

The problems that conservationists have to deal with are many and varied. Every time a house is built or a road widened to serve perhaps an airfield or even the Channel Tunnel, a part of the living world must die. A farmer sprays his field with weed-killer and a dozen flowering plants are gone. The oakwoods are threatened because timber merchants prefer the faster growing fir trees under which little vegetation will grow. Troops from the Pony Club ride across the sand dunes where the Nottingham Catchfly and the Burnet Rose struggle to survive. Hikers wear down the paths, or worse still, don't keep to them. Farmers pull down the hedges, drain the marshes, plough up the chalk downs, and blanket the fields with alien grass.

On these and other issues conservationists have learnt to compromise. It is difficult to justify the expense of a nature reserve which no one is allowed to visit. If conservationists fought all the battles, they would earn a reputation for being unreasonable and out of touch with the world. So only the burning issues are taken up and, even then, negotiation rather than confrontation is the custom.

Good relations are cultivated, not only with the general public, but with planners, farmers, water authorities, schools, anglers and ramblers.

And the reserves? Some of them are opened without restriction; in others visitors must keep to the paths; for some a permit is necessary. Others are closed at certain times of the year. Only a few are completely closed – at least in theory.

There are reserves and reserves. The official Government agency, the Nature Conservancy Council, set up in 1973, is responsible for selecting, designating and managing the national nature reserves. The Council can buy as much land as the owners will sell or its funds will allow. At the time of writing there are around 200 national nature reserves in England, Scotland and Wales. Nearly one quarter of them are owned by the

Nature Conservancy Council, and the balance are part owned or leased, or managed under a Nature Reserve Agreement.

The national nature reserves have not been set up primarily to protect rare plants or for that matter rare animals. The aim is 'to preserve and maintain as part of the nation's natural heritage, places which can be regarded as reservoirs for the main types of community and kinds of wild plants and animals represented in this country, both common and rare, typical and unusual, as well as places which contain physical features of special or outstanding interest'.

The emphasis has been on communities of plants, rare or common, rather than on individual rarities, and on occasions where a particularly rare plant grows in isolation, it is protected in many cases by local Conservation Trusts (of which, more later).

The national nature reserves are, however, quite distinct from the national parks established in such localities as Exmoor, Dartmoor, the Brecon Beacons and the Peak District. These are not nature reserves but areas of scenic beauty for which the Countryside Commission is responsible. This body seeks to protect and enhance the countryside and to provide for public enjoyment and access to it, working closely with voluntary bodies, the Royal Society for Nature Conservation, the Royal Society for the Protection of Birds, the local authorities and all concerned with the maintenance of footpaths, bridleways, and rights of way. The Countryside Commission is responsible not only for designating the national parks and the 'areas of outstanding natural beauty', but also for bringing together all those who are looking after the countryside and interpreting their views. On the whole, the emphasis has been on conservation rather than access, since without conservation there would be nothing worth visiting. The policy of the Commission is to work with farmers in order to encourage demonstration farms, tree planting (of the right kind) and coppicing – so beneficial to flowers and certain types of wild life – and to encourage the government of the day to find more funds for conservation projects.

The essential principles of a Code recommended by the Countryside Commission are:

1. Drive slowly on narrow country roads.
2. Use gates and styles to cross fences, hedges or walls.
3. Avoid damage to all boundaries.
4. Fasten all gates.
5. Keep to the public footpath over farmland.

6. Keep dogs under proper control.
7. Leave livestock, crops or machinery alone.
8. Guard against risks of fire from cigarette ends, sparks or embers from picnic fires etc.
9. Help to keep all water clean.
10. Leave no litter (especially bottles which, if broken, can injure livestock; by concentrating the rays of the sun, glass can start fires.)

Scotland has its own Countryside Commission.

Relations between the conservationists and the farming community have at times been strained, but experience shows that when planning to walk across farmland it almost always pays to ask for prior permission – if one can find out locally who is the right person – before leaving the high road. (One hint that does not appear in the official Country Code is some advice on climbing a locked farm gate. Do so at the hinge side where your weight will have minimum effect, and not at the latch side). It is also worth remembering that even the most placid and inoffensive cows will bridle at the sight of a dog. If you don't have time to get leave in writing to cross private land, it is worth going to see the farmer. The chances are that you will be shown rarities that you hardly dared to look for. In Scotland a key borrowed from the factor to a gate that straddles a highland road can let through a car and save the pilgrim botanists three or four hours of walking. In Scotland and parts of Yorkshire special care is needed at the time when the highland flowers are at their best to avoid disturbing the grouse. For instance at the Devil's Elbow to the south of Braemar, visitors who want to walk over the hills to Caenlochan Glen are asked to keep as close as possible to the pylons of the overhead skiers' chairlift. Equally in the copses of the south pheasants will be nesting about the time that the spring flowers are at their best, though at other times there may be no objection to a walk in the woods.

Apart from the Countryside Commission there is the National Trust for Places of Historic Interest or Natural Beauty, set up in 1894, which is of some interest to botanists. Although its principal aims are to maintain and enhance historic houses and buildings and to preserve landscapes of special beauty, in a number of cases – as for instance on the Lizard Peninsula – land owned by the Trust shelters botanical rarities. Some nature reserves are owned by the National Trust and can never be sold, mortgaged, nor compulsorily acquired without the special Bill of Parliament. In many cases the Trust's land is farmed, or in the case of houses and gardens occupied privately or let to tenants. In other cases, however, the properties are not owned by the Trust but are protected by restrictive covenants which the owners have given to the Trust or which the Trust has bought. These covenants continue despite changes of ownership but the land or buildings are not proof against compulsory acquisition and visitors have no rights of admission to covenanted land or buildings. The National Trust covers England, Wales and Northern Ireland. The National Trust for Scotland, is a separate body. Membership of the National Trust for Scotland allows free admission to its own properties and those of the National Trust in England, Wales and Northern Ireland which enjoys reciprocal privileges in Scotland.

Reverting, at this point, to the work of the Nature Conservancy Council, we should mention here that in addition to declaring and managing the national nature reserves, the NCC has the duty of notifying the local planning authority and the landowners concerned of areas not already in a nature reserve if they consider that the land is of special interest by reason of its flora, fauna or geological or physiographical features. Some 4,000 of these 'Sites of special scientific Interest' (SSSI) have been declared. The owner of the land designated in this way is precluded from carrying out any operation which might damage the site without first entering into negotiations with the NCC. Understandably, SSSIs vary in size. For instance, among the fifty or so sites in the Peak District National Park are small ones in railway cuttings measuring less than a hectare, whereas the Kinder-Bleaklow Plateau site is 11,000 times as big. Of the SSSIs (known colloquially as 'triple S eyes') some have been termed key sites, being considered to be of national or international importance.

Apart from the sites designated by the NCC, there are those declared by a local authority in consultation with the NCC under the National Parks and Access to the Countryside Act of 1949 as amended in 1973. This power extends from the regional councils right down through the county and district councils to the borough councils. But the vast majority of sites are those set up independently by local conservation trusts which in turn are associated with the body of which Prince Charles has become patron – The Royal Society for Nature Conservation, with a membership of more than 140,000. The local trusts and the RSNC manage, between them, 1,300 or more nature reserves in the United Kingdom extending over more than 170 square kilometres of countryside. In its own words: 'The RSNC advises and gives assistance to the Trusts on a wide range of problems; it represents their interests at national level and supports their close liaison with other voluntary bodies and with statutory bodies, particularly the Nature Conservancy Council. Together the Council and the Trusts give information and advice to local authorities, regional water authorities and other official bodies, and to landowners and farmers about the management of wild life habitats in the countryside generally. In many counties special arrangements have been made for the protection of roadside verges, rivers and streams, disused canals and railway tracks, and ancient hedges – often contributing to the protection of rare or threatened plants thereby.'

A go-ahead county organisation such as the Suffolk Trust for Nature Conservation has for example 30 major reserves and more than 80 roadside verge nature reserves under its care.

Disused railways form impressive sanctuaries for rare flowers. The Yorkshire Naturalists' Trust Reserve at Salt Lake Quarry near Horton-in-Ribblesdale, a shaded well-watered, split-level green bank with all kinds of nice orchids and other specialities abuts on to a railway property. (You need a permit to visit it.) And further east in Yorkshire two reserves have been set up on the old railway line running between Beverley and Market Weighton near Goole. We must be thankful that such railway embankments are usually too steep to be ploughed up and are seldom sprayed with weedkiller.

The RSNC also publishes books and pamphlets on nature conservation topics and produces *Natural World*, a lavishly illustrated magazine for trust members three times a year. The Royal Society for Nature Conservation assists local nature reserves, either with direct advice, or through the society's magazine *Natural World* which records the achievements and experiences of local trusts countrywide, as for example, the successes of some county trusts in promoting new reserves by purchase and lease of management agreements. Photographs show what can be done in worked-out quarries and other unorthodox sites.

All reserves need fine tuning of one kind or another – even if this is confined to a few direction signposts at the entrance. But no single treatment suits every case. Woodland reserves, in which many of our rarest plants are to be found, are as problem-racked as any because of the difficulty in arranging the right amount of canopy shading to suit individual species. Some woods require regular coppicing to let in the light. *Cephalanthera rubra*, Red Helleborine, apparently requires a half-and-half environment with some light and some shade. *Epipogium aphyllum*, the Ghost Orchid, prefers, as one might expect, deep, almost *impenetrable, shade. Lobelia urens*, the bright blue Heath Lobelia, flourishes after the woods in which it often grows have been coppiced. The reserve warden must also be something of a game-keeper. Foxes, on the whole, are to be encouraged since they keep down woodmice which have a voracious appetite for flower seeds. Insects of certain kinds – including butterflies and moths – must be present if the flowers are to be pollinated. Slugs, on the other hand, are to be exterminated: they have been found guilty, often in their absence, of feeding on the stalks of orchids.

In general the reserves must not be allowed to become too 'natural'. In some reserves the deer must be culled; in others sheep introduced to keep the grass short. In one, river reeds are harvested and sold in order to clear the mainstream. Elsewhere, trees are planted to replace those that have been felled. Grants are made to farmers who agree not to 'improve' their meadows with fertiliser so that they can buy fodder elsewhere to make up for the loss. (But more than one farmer believes that the unimproved meadows with a wealth of different species may improve the health of livestock and save the vet's bills since the tap-rooted plants permanently installed on this sort of land bring up useful chemicals which the shallow-rooted seeded grass would never reach.)

Not all rare plants are safely protected in nature reserves. But they do nevertheless enjoy back-up protection. The Conservation of Wild Creatures and Wild Plants Act of 1975 placed a ban on the uprooting of *any* plant without the permission of the landowner or occupier. It also provided a List of Scheduled Plants, no part of which might even be picked, uprooted or sold, still less destroyed, by the unauthorised 'unless the picking, uprooting or destruction occurs as an incidental result, which could not reasonably have been avoided, of any operation which was carried out in accordance with good agricultural or forestry practice'. In 1981, however, the List of Scheduled Plants was increased from 21 to 62 and the fine was upped from £100 to £500 for each plant affected. The maximum penalty is now £1,000. The List of Scheduled Plants is reviewed and revised every five years (see p. 167).

In the case of some decorative wild plants, however, a practical way of protecting them is to propagate them in quantity for commercial distribution so that they can be sold to gardeners of all kinds through plant centres.

Where, however, the plant is not especially attractive to gardeners, but is nevertheless likely to become extinct, two courses are open to the conservationist. One is to 'translocate' some of the plants from their existing site to one that is more favourable. This was the course adopted to save *Schoenus ferrugineus*, Brown Bog-Rush – with unexpected results – when the North of Scotland Hydro-Electric Board decided in 1945 to raise the level of Loch Tummel on Tayside by seventeen feet. The unforeseen consequence was that one of the botanists concerned, Mr J.A. Whelan, took one of the plants out of the vice-county in which it grew and transferred it to Loch a Choire, near Killiecrankie, which was in a different vice-county, thus upsetting past and future natural records for the plant. The original plants which had been moved up to the new higher level above Loch Tummel were destroyed by the giant waves served up by the enlarged loch. The only rushes to survive were the plants that had been taken away – one to Cambridge and the other to Loch a Choire, where it proceeded to grow larger and more luxuriant season by season.

Only one thing would have concerned the purists still more deeply: the recording in a vice-county of a species previously unknown there, followed by the discovery that it had been deliberately planted there without the knowledge of the official recorder. This brings us to the subject of deliberate introductions and re-introductions. Apothecaries, farmers, game-keepers, and landscape artists have flooded the countryside with exotic plants for perfectly good reasons. *Symphoricarpus rivularis*, the Snowberry, from North America was planted during the nineteenth century as a shrub which would give cover to game. *Carpobrotus edulis*, one of the Mesembryanthemum family and sometimes known as the Hottentot Fig, was released to beautify the cliffs of Devon and Cornwall (even though it now sometimes smothers everything else). *Lupinus nootkatensis*, the dark blue Nootka Lupin from north-west America, looks equally decorative after its fashion on the banks of the Dee. The blue-flowered chicory was originally cultivated by farmers as a crop for feeding cattle.

But there can be mishaps – just as in the case of *Polygonum cuspidatum*, Japanese Knotweed, which has become far too invasive for our comfort, and the Council of Europe's Convention on the Conservation of European Wildlife and Natural Habitats calls on its signatories to maintain strict control over the introduction of such non-native species.

But if the deliberate introduction of aliens is considered as a distortion of the environment, what feelings should we have about the re-introduction of species which have become extinct, and for that matter to the re-stocking of species the population of which is rapidly dwindling? In this field the environmentalists believe it is desirable to re-stock only when a species has become rare because of man's activities and is unlikely to be able to re-establish itself through natural regeneration. Artificial regeneration through re-stocking should take place under the same conditions and in the same environment as would govern a natural increase in the numbers of the species in question. Ideally, if seed is used it should be obtained from the existing plants. This was the method for re-stocking *Saxifraga caespitosa*, Tufted Saxifrage, which, at one time, had been reduced to a population of four plants only on a single ledge in Snowdonia measuring 3 cm x 10 cm. In some cases seed banks of increasingly rare plants have been built up under the auspices of the Royal Botanic Gardens at Wakehurst Park in Sussex. Re-introduced plants can take the form of mature plants or seedlings grown under cover and hardened off, before implantation in soil typical of the environment.

— 4 —

Recognising Rarities

Botanists who like to be 'in the know' will have become members of the Botanical Society of the British Isles. This society has an international clientèle drawn from universities and academies worldwide, and in Britain its members undertake surveys, often in collaboration with the Nature Conservancy Council, of particular plant species and groups of species. Its members also combined to provide the information needed for the monumental *Atlas of British Flora* (1962). The society can trace its ancestry back to 1836 when the Botanical Society of London was formed. In 1858 this evolved into the somewhat academically orientated Botanical Exchange Club whose members assembled dried specimens of plants to be sent to other members. George Claridge Druce who was to become one of the most famous of Britain's field botanists became Secretary of the Botanical Exchange Club in 1903 and retained the office until 1932, the year of his death. Under his patronage amateur botanists of many different hues – aristocratic dilettantes, political 'green party' reformists and untutored enthusiasts – swelled the membership much to the distaste of the élite botanists. Druce died a rich man from the proceeds of the pharmacy he set up in Oxford in 1879. It provided among other requisites 'purple specials' renowned for dispelling the hangovers suffered by the undergraduates. He published floras of Oxfordshire, Berkshire, Buckinghamshire and Northamptonshire which are still eagerly sought after by collectors. His *Comital Flora of the British Isles* was published in 1933. A generation of botanists consulted it when they wanted to know which plants they might expect to find in which counties, though senior botanists found it so full of mistakes as to be useless. Nevertheless David McClintock in his enthralling *Companion to Flowers* (1966) cites an example of Druce at his best. One autumn day in 1904 Druce came across a dried-up specimen of a grass – part of a collection that had been made in 1726 by the German botanist John Jacob Dillenius. The specimen was without a label but Druce recognised it at once as something unusual. Could it, perhaps, be the grass that Dillenius had named as *Koeleria vallesiana*? It had not been seen since and was not known to grow this side of the Loire Valley. Yet a handwritten note said that this specimen might have been found on Brean Down, a narrow finger of the coastline stretching out into the Bristol Channel to the south of Weston-super-Mare. So what did Druce do: that very afternoon he took the train from Oxford to Weston-super-Mare, got into a cab and proceeded to look around him. And there, just as Dillenius had said, was the very grass that botanists in Britain had given up hoping for. It was growing plentifully yet dozens, perhaps hundreds of botanists had passed it by unknowingly. So Dillenius really did deserve his reputation. With hindsight one can detect a slight difference in the lower part of the stem between Dillenius's Grass and the much more common Crested Hair Grass.

But we have come a long way from Druce whose name is commemorated in *Thymus drucei*. The honour came to him because he was the first to describe and publish officially the differences – mainly concerning the position of the hairs on the stem – which distinguish the Common Wild Thyme from the larger 'pop-up' species *Thymus pulegioides* and the humbler *Thymus serpyllum* known, because of the localities it favours, as Breckland Thyme.

The Botanical Exchange Club became the Botanical Society of the British Isles (BSBI) soon after the end of the Second World War and remains the principal botanical society of the country. The Society is for amateurs and professionals and it publishes a six-monthly magazine, *Watsonia*, which reports new developments in the naming of flowering plants and ferns of the British Isles and details of distribution including new records. It also provides a panel of referees should disputes arise as to the identity of plants. It has a Conservation Committee which keeps an eye on endangered plants, and co-operates with other organisations to meet threats to such places as the Somerset Levels and Amberley Wild Brooks from 'development', pollution, intensive farming and the like. The Society holds field meetings every summer – some limited to small numbers in order to protect a habitat or to make a group more manageable.

The long-standing inability of the authorities to retain the same administrative county boundaries has been a source of much distress to botanists. Rutland and Westmorland have been absorbed, Aberdeenshire, Morayshire and Banff have turned into Grampian, Argyll has been swallowed up by Strathclyde, and Perthshire has become Tayside.

Humberside has been created from Yorkshire and Lincolnshire, Hereford and Worcester are as one. No more needs to be said – except that none of these boundaries coincide with those according to which botanists have traditionally kept their records.

These were laid down by a Yorkshireman, Hewett Watson (1804–1881), who inherited an estate in Derbyshire, but preferred to live for the last forty-eight years of his life in Thames Ditton. He was a phrenologist by inclination and edited the *Phrenological Journal* for some years. But his value to botanists was the series of systematic works that he produced recording the distribution of plants in Great Britain. For this purpose he divided the country into 112 areas, known as vice-counties, with 40 more for all Ireland. Most of them were created by dividing the existing counties – Yorkshire was split into five parts – and Lothian, like Caesar's Gaul, into three. Subsequent records and much of the work of the present BSBI recorders is based on Watson's vice-counties. There was a separate vice-county (No. 68) for Cheviot and 'South Ebudes', 'Mid-Ebudes' and 'North Ebudes', representing Islay, Mull and Skye respectively.

The BSBI still bases its recordings on vice-counties first because they are more easily managed by recorders than the new larger units, secondly because new records are more easily comparable with those of the past, thirdly because when writing about plants one can be more specific if one refers to smaller areas, and finally because 'they' may yet change the present county boundaries once again.

The Plant Records, included regularly in the Society's magazine *Watsonia*, are greatly encouraging. They include, for each vice-county, finds made for the first or second time since 1930 or instances where the range of the plant has been extended by more than 100 km. Thus *Myrrhis odorata*, Sweet Cicely, a plant of the Cow Parsley family is common enough in the north of England, but when found at Stokehill Farm, North Hampshire, in 1983 represented the second record only for that vice-county. A Wild Raspberry found in Roxburgh was the first localised record since 1887.

Watsonia nevertheless also contains electrifying accounts of the plants recorded on BSBI botanical expeditions, that lead one to wish that it had been possible to go on all of them whatever the weather.

These forays are not just fun sessions. A recent expedition to Wester Ross for instance involved visits to two national nature reserves, including limestone pavements, limestone gorge woodlands, uplands and bogs with accommodation in a bothy (built for farm labourers) holding up to six, or on a camp site. Their purpose is to explore little worked areas in order to study a particular group of plants or to help in preparing or revising local *Floras*. They are not, the Society points out, 'for the purpose of conducting members to the localities of very rare plants', but rather to areas which promise an interesting, varied and unusual flora. But clearly it is from such expeditions to likely spots that discoveries of fresh rarities could emerge.

The BSBI is currently conducting a new survey throughout the British Isles of sample areas within each 10-kilometre square to record which species have undergone changes in frequency or distribution since the publication of *The Atlas of the British Flora* in 1962.

Another admirable source of knowledge is the Wild Flower Society which celebrated its centenary in 1986. Its aims are to promote a greater knowledge of field botany among the general public, to advance education in matters relating to the conservation of wild flowers and the countryside, and to promote in every way the conservation of the British flora.

Membership is open to all who support the aims of the society. Junior membership and group membership are available at reduced subscriptions. Any publications the society may produce are available on payment to non-members, and guests are welcome at field meetings on the same condition as members. Their *Wild Flower Magazine* is published three times a year.

Members may, if they wish, record their finds in a field botanist's book known as the *Wild Flower Diary* (see p. 170).

— 5 —

Planning Expeditions

'Excursions may be truly said to be the life of the botanist,' as a famous Scottish professor, John Hutton Balfour, once declared.

Much time is saved by joining group expeditions with a leader who already knows where to look for interesting plants. Nevertheless there is no thrill quite like discovering a rarity unaided. However, great caution is needed. Even the lone mountaineer, undaunted by the vertical cliffs of Snowdon's Clogwyn Du'r Arddu, can get into trouble, and the lone scrambler is probably even more vulnerable. It is a solemn thought, or it should be so, that when, in order to reach firm ground, you slither perhaps only three feet down a rock-face, there may be no way of getting back to the spot from where you started. You may have to go on – and on – and who knows where that can lead? Such a dilemma can arise very suddenly for the climber who is making a traverse along the flank of a mountain from one ridge to another without being able to see exactly what lies beyond the next brow.

The second danger to the lone novice on the mountainside comes from mist or fog. It can take over with surprising speed, blotting out not only landmarks and peaks above, but precipices below – distorting, incidentally, your calls for help – if you have not remembered to carry a whistle on a lanyard round your neck. It is also surprising how quickly the temperature can drop in these circumstances the moment the sun goes behind a cloud.

So tell the staff of the hotel where you are staying, or the mountain rescue centre, if there is one, of your plans for the day. In some areas the mountain rescue centre is equipped with a helicopter which makes a routine patrol over the terrain even on fine days to take care of the visitor who may have twisted an ankle or even broken a leg.

KNOWING WHERE TO LOOK

To field botanists, maps can be as enthralling as a chart from Treasure

Island. To be told that some rare plant grows in a locality with a particular map reference is to be faced with a challenge that must be answered. Full information on the kinds of detailed maps available and how to read them are to be found in the Appendix on p. 172.

Though it is helpful to study geological maps in advance, once on the ground, most botanists can tell whether they are on what is called basic or alkaline soil by the flowers that grow on it – wild thyme, cowslips, some bell-flowers, gentians and the like. Similarly, acid soil yields the heather, Tormentil, Bilberry, Wood Sage and the inelegantly named Lousewort.

As to rocks: granite of which we see so much in Scotland is relatively unproductive, but along with it is found a mineral stone on which superb rarities can often be found. This is serpentine: composed of magnesium silicate plus a little ferrous iron and aluminium. It is green in colour but frequently mottled with red like a serpent's skin – and in fact, a reddish stain on the side of the cliff is the clearest sign of its presence.

It is seen all too seldom in the glens of Scotland, but wherever it appears one can be certain of finding treasures. John Raven in the book on mountain flowers which he wrote with Max Walters has described how he went to see the rare Alpine Catchfly on the treacherous Hobcarton Crag in Cumbria. He did his best to describe which of the several gulleys one should clamber up in the hope of finding the plant. But the only one worth climbing is the one with a red stain at the bottom. There have been suggestions that the mineral in question was the iron sulphide known as pyrites. But certainly the only other place where the Alpine Catchfly is to be found is on serpentine rock above Glen Doll.

Apart from this, serpentine rock is to be found on the Lizard Peninsula. There are fine flowers to be found there, some of which are not to be seen elsewhere. At Kynance Cove, serpentine rises in columns and stacks from the sand, coloured not only green and red but blue and purple.

Finally, forgetting all rarities for a moment, there is the truism that travel broadens the mind. Thus one of the exciting discoveries for a traveller who lives in the south of Britain is that, north of the Trent, plants that he has never seen and is unlikely ever to see in his home county are present in abundance. He is filled with admiration for the delicate leaves of *Myrrhis odorata*, Sweet Cicely, divided, subdivided and then slashed still further, and because the plant grows beside every other stone wall, he is not ashamed to pluck off a leaf in order to confirm that when crushed it does indeed recall the scent of the aniseed gobstoppers of schooldays. Then there is *Geranium sylvaticum*, Wood Cranesbill, with its sharply tinted azure-mauve petals, which conveniently flanks the pathways along the main rides of the forest. Similarly travellers from the north might be surprised to find the magnificent Tree Mallow, *Lavateria arborea*, and the vivid Red Valerian, *Centranthus ruber*, so firmly established along parts of the south coast. Yet not one of these four plants deserves to be called rare or in need of protection.

Books, even if published some years ago, are not to be despised by the plant hunter who is planning an expedition. *Finding Wild Flowers* by R.S. Fitter (1971) gives an excellent summary, county by county, of the specialities growing in each, with mini-maps. *Mountain Flowers* by John Raven and Max Walters (1956) provides all the thrill of rock-climbing for plants – even for those who may not have the time or the energy to drag themselves uphill. *Wild Flowers of the Chalk and Limestone* by the (late) J.E. ('Ted') Lousley (1969) gives the reader a gentler ride across the downs and dales, while V.S. Summerhayes provides a specialist's view in *Wild Orchids of Britain* (1985) and gives more of the background than some recent works on the same subject. Then, quite lately, there has been a flood of beautifully produced County Floras – such as *The Atlas of the Kent Flora* published by Kent Field Club and *The Flora of Surrey* by J.E. Lousley. *The Flora of Jersey* from the Société Jersiaise of St Helier deals with more than 1,500 species, with distribution maps for nearly 600 and 18 colour plates. And these are but a few examples of many: new Floras keep appearing almost month by month like bean-sprouts.

Among the aids to identification which can conveniently be carried in a haversack, the best known is probably *Collins Pocket Guide to Wild Flowers* by David McClintock and R.S. Fitter. This has the advantage of coloured illustrations, as has *The Wild Flowers of Britain and Northern Europe* by Richard Fitter, Alastair Fitter and Marjorie Blamey. *The Wild Flower Key* by Francis Rose is another extremely valuable guide, well-illustrated, with plenty of background information on each species.

In this book we have tried to avoid those flowers at risk which are hard to identify, and have turned rather to those which can be recognised and loved at first sight, and perhaps photographed. However, accounts in detail are sometimes indispensable, and for this purpose the standard work – not for the pocket in any sense of the word – is *Flora of the British Isles* by A.R. Clapham, T.G. Tutin, and E.F. Warburg. The second edition of this 1,269 page work was published in 1962 and another is promised soon. *The Excursion Flora* – in essence an abridged form of the above – is highly recommended as a portable guide but is to my mind disappointingly cursory in its treatment of the more interesting rarer plants.

A book that most nature-lovers would love to be able to afford is *The Macmillan Guide to Nature Reserves*, researched and written by Jeremy Hywel-Davies and Valerie Thom. This is not only well-written, attractively presented with district maps, and in some cases the warden's telephone number, but it is also decorated with numerous colour photos of plants and habitats showing the would-be visitor what to expect. Understandably there is some reticence as to the location of rare plants and some of the rarest are not even mentioned. But for a good read in winter and for planning visits to the places one is unlikely to have enough time ever to visit, this book is almost ideal though, with new reserves being extended or set up almost monthly, it will have eventually to be updated. (Since going to press, a cheaper paperback edition has appeared.)

A HUNDRED
ENDANGERED PLANTS

— I —

Among the Rarest

* (See facing pp. 66 and 67 for illustrations)

Ia. *Gagea bohemica* (no vernacular name)

Is it not astonishing that the flower of a species unknown to Britain should have been blossoming in this country for perhaps five thousand years or more without having been observed? Yet this is the case with the dainty yellow-starred *Gagea bohemica*, discovered so recently here that no one has yet devised an appropriate name for it in English.

The stems, with usually but a single flower at the end, are between 1.5 and 3 cms long (say ⅝–1¼ ins). The leaves on the flower stalk are in pairs, opposite to one another, thin and strap-like. Those from the base of the flowering shoot are thread-like and nearly three times as long as the flower stem and are disposed round about it in a dishevelled manner. The flower varies in diameter between $\frac{1}{10}$ in and $\frac{1}{5}$ in and has usually six petals of a bright shiny yellow within and a greenish tinge on the underside.

In April 1965 Mr R.F.O. Kemp, a distinguished botanist – currently a lecturer at the University of Edinburgh – was collecting a specimen of some moss from the rocks of Stanner (the same locality in which *Lychnis viscaria* grows), and was surprised to find that he had apparently seized with it a small white flower. After some deliberation he recorded it as *Lloydia serotina*, the Snowdon Lily. Six years later, also in April, Mr R.G. Woods, of the Nature Conservancy Council, visited the same spot, and found a further flower, also white, and somewhat shrivelled. A comparison between this plant and a photograph of Mr Kemp's flower showed they were of the same species. But neither of them was *Lloydia*, because both the photograph, and Mr Woods' plant showed traces of hair, whereas *Lloydia* has none.

The mystery was solved when Mr Woods paid a second visit to Stanner Rocks in mid-January 1975 and found a plant in full bloom. Its flowers were yellow and would soon have faded to white, when they would have been a replica of what Mr Kemp and Mr Woods had seen in April. Indeed in that year the petals had already turned to white before the end of February. Two experts, J.A. and J.H. Schultes, identified it tentatively as *Gagea bohemica* – a courageous proposition since it had never before been seen in Britain – but their judgement was confirmed in 1978 when David McClintock and E.M. Rix met R.G. Woods at Stanner Rocks and found 25 specimens in full flower.

So far, David McClintock has not come up with a suggestion for an English name as apt as the one he promoted for the Ghost Orchid. What should it be? Stanner Starlet? Early Gagea? Welsh Gagea? The field is open.

One would have thought, too, that a new botanical name might also be appropriate. Certainly the plant does grow in Bohemia. But it has also been found in the valleys of the Loire, the Seine, the Rhone; in Sicily, Corsica, Sardinia, Greece, Turkey, Israel and the Soviet Union and doubtless other places to be announced. So why should Bohemia have the monopoly?

It belongs to the Lily-Tulip-Fritillary group but is not really like enough to any of these to be linked with them. The genus Gagea was named after Sir Thomas Gage, 7th Baronet of Hengrave Hall, Suffolk (1781–1820) who, as David McClintock recorded in his fascinating book *Companion to Flowers*, nearly lost his life while climbing in Snowdonia in search of *Lloydia*. A fair enough reward one might suppose, though if every unlucky botanist was to be commemorated in this way we should soon have enough names for a telephone directory.

Ib. *Helianthemum apenninum* White Rock-rose

* 6th June North Somerset

Here is a modest plant which has helped to give rise to many of the more garish varieties seen in gardens. The flowers are milk-white and the leaves, as if to match, are covered with greyish short cottony hair on both the upper and lower surfaces. The edges of the leaves are often 'revolute', that is, turned under. The sepals, like the leaves, are also covered with greyish cottony hair.

Rock-roses are especially generous with pollen and the flowers are visited by various pollen-eating insects. The stamens are described botanically as 'irritable', and when an insect, lands they spring outwards from the centre of the flower towards the petals. The purpose of this manoeuvre is not at once apparent. It could be a mechanism for avoiding self-fertilisation, or possibly a way of ensuring that not all the pollen is used up by one visitor. Or – more likely – the arrangement displays the anthers to better advantage, allowing the visitor to reach more of them with greater ease. It is also believed that some of the generous supplies of pollen are wind-blown and that some fertilisation takes place in this way. In the wild the plant is to be found on limestone in half a dozen or more sites, all in either Devon or Somerset. The best known location – referred to in many books – is on Brean Down, a grassy peninsula to the south of Weston-super-Mare. It overlooks the Bristol Channel with a view on a clear day over to Lundy Island where the naturalised/wild Paeony grows on the slopes beneath a former priory. Brean Down is a small area and an easy climb, yet for the most part the plants seem to remain undisturbed, perhaps because they are scattered and do not offer a massive display or even recognisable colonies. Besides, a frail little semi-creeping stem and a relatively delicate-looking flower is not going to last long in water. White rockroses are often to be found at the nearest nursery and one particularly reliable firm describes a Helianthemum with grey foliage and white flowers as 'The Bride'.

Ic. *Oxytropis campestris* Yellow Oxytropis

* 13th July Tayside

Few members of the peaflower family – to which this belongs – can rival this one for beauty. Not even the Sea Pea with its roseate bunches of flowers, nor the spires of the elegantly pink striped Sainfoin, nor the exuberant Wood Vetch smothering the bushes with white blossoms veined purple as if with indelible pencil.

It is surely the mass of delicate silvery leaves and the bouquet of ivory (I would have thought, rather than yellow) blooms which it offers on the end of a green stalk decorated with a frosting of silvery hairs that sets this plant apart.

It was discovered back in 1812 by the remarkable Scottish botanist George Don who began to explore the highlands in 1779 and was the first plant hunter to investigate Glen Clova. It was also one of the flowers that John Hutton Balfour (1808–1844), Emeritus Professor of Botany at Edinburgh, singled out for praise when he wrote of expeditions which, he said 'may truly be said to be the *life* of the botanist'. To his knowledge – he was writing in 1840 – the plant confined itself 'to a single British cliff'.

Almost certainly this is the one, a scrap of which is to be seen in the photograph f. p. 66 and described in greater detail in that memorable book *Mountain Flowers* by John Raven and Max Walters. If so, it is a climb not lightly to be undertaken, and this photographer, having reached the summit found himself looking down on the propellers of the yellow Mountain Rescue helicopter on its diurnal search for benighted botanists. How nice – but how expensive – it might have been to have beckoned it over for a quick descent, instead of plodding down a sheep's traverse to the foot of the corrie. A direct climb up would, I imagine, be inadvisable for the area below the plant, distinguished by its ochre coloration, has been smoothed into a no-hold area by the rains and tobogganing rocks. In fact I was told that the plant itself is occasionally brought down too, to flower in the scree at the foot of the cliff.

Since Balfour's day, the number of localities in which this beautiful plant is to be found has risen to 'up to six'. Apart from Angus, a vice-county which includes the site already referred to, it is found in Tayside and Kintyre, where, it is said, wild goats have helped it to prosper by keeping down competing vegetation.

Oxytropis campestris has a cousin with purple flowers – *Oxytropis halleri* – which is almost as distinctive but is less scarce, occurring in Wigtownshire and Argyll in the west and Ross and Sutherland in the north. The ascent of Ben-y-Vrackie, its haunt in Perthshire, is considerably easier than that of the cliff frequented by its cousin in Angus, but is grazed intensively by sheep.

The unusual name Oxytropis is derived from a minute detail in the flower, the keel of which ends in a sharp point. *Oxus* is Greek for 'sharp' and *tropis* is the keel. Prospectors in the Highlands may find it useful to know the meanings of terms often used by map-makers, and a list is given on p. 173.

Id. *Bupleurum baldense* Small Hare's-ear

This plant is usually so small that only a print enlarged to twelve times its life size shows its structure adequately. Even then it is difficult to reconcile oneself to the fact that Small Hare's-ear is a member of the Umbelliferae or Carrot family in which the flower stalks generally arise, umbrella-fashion, at the top of the stem. 'What stem?' one wonders in the case of Small Hare's-ear. If there really is one, it is dwarfed by the shortest of turf grass, and when members of the Botanical Society of the British Isles made a count of the plants at the well-known site somewhere above the Beachy Head Lighthouse near Eastbourne, they must have felt they were counting the stitches of some impossibly fine tapestry. The minute yellow flowers with petals rolled inwards towards the centre are so concealed by green bracteoles which surround them as to be almost invisible.

In Sussex the Small Hare's-ear does grow perilously near the edge of the cliff and there have been fears that as the cliff-top falls away – as it does from time to time – the Sussex colony will disappear. On the other hand if the plants survive only because they are close to the brink, it may be because nothing else will willingly grow in a position so exposed to the winds and salt spray. If so, as the cliff crumbles, new plants of the Small Hare's-ear may be able to grow a little further inland but still in their favourite position near the cliff-top, where, incidentally they would be safe from trampling except by the most foolhardy who leave the nature trail provided by Eastbourne's local authority.

The second mainland site for Small Hare's-ear is within the Berry Head Country Park. This lies to the south of Torquay and is controlled by the local Torbay authority. This is the largest stretch in Britain of the type of limestone known – appropriately enough – as Devonian limestone. Conditions here must be rather similar to those above Beachy Head except that one is chalk and the other limestone.

So many other rare plants – White Rock-roses, Goldilocks Aster and *Ononis reclinata*, Small Restharrow – grow here that it has been suggested that this may be a small area that survived unchanged throughout the Ice Ages and that visitors to it see here something of a vanished England that exists nowhere else. Small Hare's-ear is also to be found on several of the main Channel Islands but the plants there are rather larger.

Ie. *Ophrys holoserica (fuciflora)* Late Spider Orchid

* 9th June East Kent

This is a really rough-looking orchid with a soot-brown bristly square-jawed lip, pencilled with yellow circles, and with a yellow appendage jutting out beneath like a goatee beard. At the top on each side of the lip are two 'shoulders' suggesting comparison with something quite different, for they resemble the sombre velvet wings of an old-fashioned armchair. The remaining petals are flushed red, stubby and peg-like, tapering to a point, and the sepals are of a deep pink, with perhaps here and there a hairline stripe of green. They, too, are stubby; almost egg-shaped – one of several characteristics which helps to distinguish the Late Spider from the very much more familiar *Ophrys apifera*, the Bee Orchid (which has narrower and longer sepals, longer and non-tapering pink petals rounded at the top and the appendage tucked under the lip and not projecting forwards).

Ophrys holoserica does indeed flower in mid-June, about a month later than *Ophrys sphegodes*, the Early Spider Orchid, but is very much scarcer, being confined to a few sites in East Kent. It survives at extreme risk, since it is becoming ever more difficult to protect the chalk downs and field borders which the plant favours. The sheep, which, if maintained at a density of about three per acre, would keep the downland turf in splendid condition have largely disappeared. The rabbits have vanished too following successive attacks of myxomatosis, and coarse Tor grass and Erect Brome together with other weeds have appeared instead. The introduction of milk quotas has tempted the farmers to turn downland, which was formerly grazed, into arable land, and the modern tractor is able to plough on slopes which would have defied a team of horses. Finally, nearly all of East Kent is likely to be affected in one way or another by the advent of the Channel Tunnel.

But one other peril, till recently unsuspected, confronts the Late Spider Orchid – namely death by hybridisation. Already at least one specimen has appeared, at a site previously known for its Late Spider Orchids, of what could be *Ophrys holoserica* X *Ophrys apifera*, that is a hybrid between the Late Spider Orchid and the Bee Orchid. One plant that I photographed had the petals of *O. holoserica*, like scarlet carrots, and the lip with the appendage directed forward, but the sepals were long, narrow and dagger-like as in the Bee Orchid. The lip, too, was less square and more acute than the normal labellum of a Late Spider. The plant also had that extra vigour so often seen in hybrids. Bees belonging to the genus *Eucera* pollinate both Bee and Late Spider Orchids and some orchid-lovers fear that the new hybrids may take over and supersede the original species. The verdict of the experts remains however in suspense as to whether 'my' specimen is a hybrid, some members of the jury being of the opinion that small colonies of the genuine orchid can develop appreciable variation.

If. *Cypripedium calceolus* Lady's Slipper Orchid

* 4th June Yorkshire

Is this the most beautiful of all our native flowers? Or is it just the most garish and out-of-place bloom ever to appear in a countryside of stone walls and sheep dog trials.

It may soon be too late to worry about this if the single wild specimen of the Lady's Slipper Orchid reaches its allotted span without producing offspring.

John Parkinson in his *Theatrum Botanicum* of 1640 was the first to mention the native plant, noting that it was said to grow 'in a wood called the Helkes, fast by Ingleborough' (in Yorkshire). Thirty years later, John Ray recorded it himself in his *Catalogue Plantarum Angliae* as growing at the end of Helks (sic) Wood. It was still there in 1699, but had disappeared, according to a note in Withering's *British Botany*, less than a century later.

F. Arnold Lees, (1847–1921), a close observer of the plant, according to a series of articles for the *North Western Naturalist* between 1937 and 1939: 'Our Lady's Slipper – the great Flower Prize of Craven, on record since Ray's time yet not famous only by tradition: many of the both young and old residents having made acquaintance with it upon its erratic flowerings in this or that disturbed or slipped-earth nook of scrub wood during the last fifty years! The last assurgence in three spots in and about Grass Wood and the Skirfare Valley was in 1875–81; the latest discovery in 1906 being made by the late Missie Madge Caradice when quite a girl, on a talus bank near Kettlewell, where it is still; not blooming every year but some seasons as in 1909 the clump or colony in one spot perfected six flower spikes.

'It has also occurred in living memory in Heseltine Gill, under the shadow of Penygent, in Knipe Wood, in Belmont (*Gleadall*) and in the Sleets Gill Scrub of Lower Arndale.

'It is not so much erratic as to locality, as uncertain in its flowering – the careful picking of a flower stem one year leading to a crown increase and leafy clumping for the succeeding ones until a late May or early June arrives when from reduced-foliaceous, siccate and inconspicuous crowns will rapidly shoot up several stems every one with a bloom, or even a couple at its summit. Then, amid the associated Lily of the Valley leafage it is conspicuous enough for quite a week (or more, gathered, in water).'

Lees' localities for the *Cypripedium* were all in West Yorkshire. But there have also been records from the other north country areas, e.g. Slip Gill near Rievaulx in North Yorkshire, Castle Eden Dene in Co. Durham, from Westmorland and Cumberland, and even Derbyshire. Of all these only one Yorkshire site survives today and it is occupied by a single plant.

The bright yellow lip or labellum, balloon-like and shining, as if lit from within, makes the flower distinguishable from afar. Technically speaking the lip is one of the three petals. The other two petals and the three sepals (some botanists do not distinguish between sepals and petals, calling them both 'segments of the perianth') are of a colour varying from that of vin rosé to that of vintage port. Of the outer perianth segments (i.e. those at the back of the flower) which we may regard as the sepals, the largest stands upright like a sentry box above the flower. The remaining sepals are joined together beneath the lip like Siamese twins, for almost their entire length. Only the tips remain to show that they might in earlier times have been separate. The two side petals, narrow and somewhat translucent, are often twisted into lax spirals.

The third petal, the labellum, as if to display its contours, is marked with faint lines of longitude. Looking at it from above, one can see why it has reminded botanists of a Lady's Slipper, and led to the botanical name *calceolus* meaning slipper-shaped. The opening at the top, through which one might imagine a Cinderella's foot entering, is rounded, and the edge tucked over, a feature which, as we shall see, promotes fertilisation.

In this orchid as with other orchids, the 'male' reproductive organs that produce the pollen, and the 'female' which receive it are fused into a single structure known as the column. In the Lady's Slipper, the outer (foremost) part of this column termed the staminode, is yellow, spotted red, and curves forward, from the neck of the flower, like a narrow but flamboyant tie. The two stamens – an unusual feature distinguishing the *Cypripedium* from most other orchids which have only one – are to be found one on each side of the base of the column.

In the wild state, *Cypripedium* depends for pollination on the visits of certain insects of the right size belonging to the bee genus known as *Andrena*. These are attracted by the scent of nectar which lies on the floor of the slipper-like lip of the flower.

They endeavour to land on the staminode but this offers no foothold so they fall down into the slipper. The inner surface of this is smooth and overhanging so that it is impossible for the bees to escape from the flower by the route by which they entered. Instead they make for either of two smaller openings of roughly their size at the rear of the slipper. They are helped in their climb by the stiff hairs which line the approach to the exits. On the way up their bodies rub first against the stigma on the inner side of the column and then against one of the anthers, from which some of the sticky pollen rubs off. If they then visit a second flower, the pollen is rubbed off against the stigma when they leave. However it is worth recalling the words of Arnold Lees who wrote that the flower was 'conspicuous enough and yet I never saw it visited by an insect, nor do I think, if left to fulfil its mission, it commonly ripens seed'.

Assuming, however, that fertilisation takes place, and that viable seed is produced, the next step is germination, the development which leads eventually to the appearance of new tissue in the form of roots, leaves etc.

In the case of many orchids, however – and the *Cypripedium* is one of these – development is assisted by the presence of certain fungi. This strange form of co-habitation is described as mycorrhizal because the fungi are present in 'roots', and as symbiotic because two different organisms develop together for their mutual benefit. This relationship developed either because of the primitive structure of the orchid seed, which is not provided with an endosperm or other structures containing reserve food material which the embryo can utilise to form infant roots and leaves, or perhaps, over thousands of years, because the presence of the fungus has rendered the endosperm which earlier orchids may have possessed superfluous. Whatever the true explanation, many orchids in the early stages of their development depend on fungi to supply them with sugary substances in their simplest form, and perhaps also enzymes and vitamins.

There seems to be some doubt as to whether, in the natural course of events, the orchid seed becomes infected by fungus before it germinates – or whether it is only after germination that the fungus exercises a beneficial effect.

The balance of evidence seems to be that infection takes place only after the embryo has already shown possible signs of germination – as for instance by absorption of water or an increase in size. Once the embryo starts to grow, the fungus apparently penetrates it either through hairs on the outer surface of the embryo, or at the point where the radicle or small root will eventually appear.

Some scientists believe, however, that compounds such as thiamine and nicotinic acid are released by fungi *before* invasion has taken place, and that these compounds may assist germination. Electronic micrographs show that once invasion has occurred, the filaments of the fungus form distinctive coils within the orchid cells, known as pelotons.

Some cells of the fungus eventually degenerate, and it has been suggested that these, too, constitute a food supply for the orchid, although the process by which this is achieved is unclear.

It appears nevertheless that the relationship between the plant and the fungus is under the control of the orchid and that in winter, when the plant is dormant, the fungi are absent.

A small number of orchids are dependent for the whole of their lives on nourishment from fungi. They are species that do not possess the chlorophyll that gives green colour to leaves, and enables plants to convert the energy of light into chemical energy. Other orchids, including *Cypripedium calceolus*, are dependent on fungi only during the earlier stages of their life cycle.

The effects of winters and summers and the rate of degeneration of the older parts of the rhizome are apparent, and it is therefore possible to count the number of years that have passed before the plant has sent up the first green leaves, and it is believed that in the case of the Lady's Slipper, the waiting period before a flower appears can be quite short but may be as long as sixteen years. In the Green-Winged Orchid, *Orchis morio*, a green leaf is usually produced after only one year's growth.

At the end of each year the Cypripedium leaves and the shoot, if any, will die down but, as the plant develops further, so the rhizome system extends and more shoots are sent up at different points. Usually each spike of the Lady's Slipper orchid carries only one bloom but occasionally there are two as in our photograph f. p. 66 and three or more have been known on British plants.

The grassy slope in Yorkshire on which the wild plant grows is dotted not with Lily of the Valley as Lees had described, but with *Primula farinosa*, Bird's-eye Primrose. Looking up the slope from the warden's low-profile tent pitched beneath a wall, one is impressed by the steep rise and one recalls Lees' judgement that 'the rhizomes are far-creeping and intrusive among other rootlets in an upper layer of the leafmould, shallow ever upon crevices line rock as it is, so that a sudden downpour of rain may easily remove a site, e.g. on Knipe's steep slope by a yard at least'.

One can thus well understand why visits to *C. calceolus* are discouraged at the moment. If they are ever resumed, visitors should bear in mind that they will probably be viewing the plant from a distance of at least thirty feet and that the bloom, seen through the average 50 mm lens and enlarged to a 13 cm x 9 cm print, will not greatly exceed $\frac{1}{10}$th of an inch in size. To get a recognisable photo they will need a 200 or 300 mm telephoto lens, or better still a 500 mm one. On some days a long-distance flashlight is called for. Excellent high-quality pictures are however obtainable from the Yorkshire Naturalists' Trust.

It will be appreciated from the foregoing that the structure of the flower of the Lady's Slipper precludes the possibility of self-fertilisation – except when the spike bears more than one flower. Selfed seed has been obtained from the Yorkshire plant by hand pollination, but has not proved viable. Efforts have been made for more than twenty years to find other wild plants of the Lady's Slipper which could be used for cross-pollination, and volunteers from the Yorkshire Naturalists' Trust have searched likely spots in line abreast so far without success.

The alternative of pollinating the wild plant with pollen from a plant of Lady's Slipper grown in cultivation has therefore been adopted.

Pollination of the wild plants is arranged through the Nature Conservancy Council and more than one of the Nature Conservation Trusts. It is an operation that needs careful planning as flowers protected in a garden usually open two to three weeks before wild specimens. The collection of the seed also needs careful timing. The aim is to gather the seed capsules before they open to shed the seed, usually some three and four months after pollination. This operation too is beset with risks. On one occasion the single spike of *Cypripedium calceolus* was eaten by a sheep; on another occasion a rabbit carried off the prize, and in a third year, the pods split open unexpectedly, scattering the seeds before they could be collected.

After collection, the mature seed capsules are split, and the seeds are passed through a fine sieve to remove débris; they are then placed in glass vials and dried over moisture-absorbent silica before being stored.

The seed can then be grown on in laboratory conditions usually described as *in vitro*, the implication being that the work is carried out mainly in glass test tubes.

Two alternative methods of cultivation are used. In one, the symbiotic method, a chosen fungus is associated with the seed together with chemical nutrients. In the other, no fungus is used but extra chemical nutrients, mainly sugars, are added to provide those which it is thought the fungus would have yielded.

The use of fungi to promote the germination and development of seeds was pioneered by the young French botanist, Noel Bernard from 1899 onwards to the 1920s. It was carried on by Hans Burgeff of Würzburg and further developed by J.H. Warcup (1971) and M.A. Clements (1982). The system is closely akin to what happens in the natural world, and this method has proved successful in Australia with many species including one of the Donkey Orchids, the rare *Diuris punctata* var. *alboviolacea*.

Some encouraging results have also been achieved in growing European terrestrial orchids *in vitro* with fungi, using a method described recently in the *Kew Bulletin* in an article by M.A. Clements (of the Australian National Botanic Gardens, Canberra) and Harriet Muir and Dr P.J. Cribb both of the Royal Botanic Gardens, Kew.

In this method the fungi to be used are first removed from the roots of orchids already in cultivation at Kew. The roots are washed and then examined to ensure that they contain living fungal tissue. This is then carefully removed, and examined, and grown on agar plates and those strains which show the characteristics of mycorrhizal fungi are selectively cultured, on nutrient agar.

The seeds are disinfected in sodium hypochlorite solutions and are then sown under sterile conditions. They are placed on a growing medium containing agar, sucrose, yeast extract and a number of chemical compounds of calcium, potassium, magnesium, nitrogen and phosphorus.

The agar on which the seeds have been sown is inoculated with the chosen fungus and, from then on, a close watch is kept on the development of the seed. The cultures are kept in the dark until germination occurs and then given a 16-hour day under daylight fluorescent lighting at a temperature of between 19 and 21 degrees Celsius.

If the embryo germinates, it will pass through several stages of development, five of which are easily recognised. Germination, the point at which the seed swells and breaks through its outer covering, is the first of these; in the second, thread-like outgrowths known as rhizoids will appear; in the third, signs of an immature leaf are detectable; in the fourth stage, the leaf shows green coloration due to the presence of chlorophyll, and finally the beginnings of the root itself can be seen.

At the time of writing this method has been used to germinate seeds of 23 different species of European orchids. Fourteen have been successfully grown to the seedling stage of development, and it appears that they are up to one year in advance of corresponding plants grown in the wild. Predictably perhaps the commoner and presumably more robust species have shown better results than the rarer species.

Some mortality occurs during the weaning process when the vulnerable seedlings are transferred from agar to a soil-based medium. The precise reason is unclear but possibly the larger seedlings are not compatible with the original fungus.

The Lady's Slipper seems to be associated with fungi belonging to the form-genus *Rhizoctonia* – a group which includes a large number of fungi which can be differentiated as species only if they can be induced to complete a full life cycle.

Another line of approach would be to propagate plants not from seed but from cells taken from the tip of a growing shoot or meristem as it is called. These cells contain all the potential needed for a completely new plant.

Several groups of cells can be cut from a single meristem and each could eventually develop roots and leaves of its own which can then be multiplied up into an ever-increasing number of new orchids (though in the case of *C. calceolus* not so many as would be yielded by a rose which has meristems in the axilla of almost every leaf). One difficulty with the *in vitro* propagation from meristems is that, in the case of the Lady's Slipper and other terrestrial orchids, it is not easy to carry out the necessary preliminary disinfection of the meristem without damage to the tissue.

Nevertheless, if *in vitro* propagation could be achieved using garden plants of *Cypripedium*, this would add to the security of the wild plant, since many of its admirers would address themselves to the nearest garden centre without troubling to look for the wild plant itself.

These attempts at mass production are part of a world-wide movement to save orchids.

For other countries apparently far more richly endowed with orchids than we in the UK have even greater problems in saving them. Orchids in Central and Latin America are at risk from collection by hobbyists and botanists, of destruction and clearance of forests for timber and/or agriculture, from mining and from projects for hydro-electric dams and new roads.

Rare orchids are being collected and sold in large quantities and it has been therefore necessary for the Convention on International Trade in Endangered Species of Fauna and Flora (CITES for short) which has been ratified at the time of writing by 91 countries and which came into force in 1974, to include all orchids on the list known as Appendix II, even those as yet unnamed. This Appendix covers 'species threatened in the wild which may become endangered through trade' and international trade in plants listed in Appendix II is controlled and monitored by licensing.

In 1984 the International Union for Conservation of Nature and Nature Resources officially took orchids under its protection by establishing a special Group for orchids under the auspices of its Species Survival Commission.

Eric Hagstater, who has been studying the orchids of Mexico and Central and South America for more than twenty years, was invited to become Chairman of the Group, and Joyce Stewart, Sainsbury Orchid Fellow at the Royal Botanic Gardens, Kew, who has been involved with the orchids of Africa for a similar period, agreed to become its Secretary.

As a start the orchid group will have the following goals:

Conservation of Orchids

To ensure the survival and maintenance of genetic diversity of all orchids throughout the world, both in the wild and in cultivation.

Data Collection

1. To establish checklists of orchid species occurring in the wild for all countries and to determine the degree of threat facing each species from either habitat destruction, or collection for trade, or both.
2. To prepare a record of important areas that are already legally protected, e.g. National Parks and Protected Areas, and lists of the orchid species that are, in theory, safe within them.
3. To establish lists of areas of great species diversity or high endemicity that should be protected by conservation organisations, governmental or otherwise.
4. To prepare a record of orchid species that are already in cultivation whether in botanical gardens or in private or commercial collections.
5. To prepare a record of those species that are already being propagated and successfully grown in cultivation.

6. To establish lists of species that should be propagated: those which are of 'value' in any sense; those which could take the pressure of wild populations; those which could be used to increase the wild populations where these are small.

Research

To promote and support much more basic research on the Orchidaceae than is currently being undertaken. Such research should include basic taxonomy and ecological, horticultural, biological and biochemical studies.

Public Awareness of the Need for Orchid Conservation

To develop an awareness of plant species and habitat conservation through published books, scientific and popular articles, films and other visual media, newsletters, symposia and meetings and their proceedings.

Develop and Promote Conservation Strategies and Special Projects

1. To promote the conservation of areas of great species diversity or high endemicity.
2. To prepare and circulate as widely as possible a 'Code of Conduct for orchid growers and collectors' whose main aim will be the further protection of species in their wild habitats.
3. To review the operation of CITES with respect to the international trade in orchid species and recommend changes that would make its implementation more successful and more effective for conservation.
4. To encourage the artificial propagation of orchid species, especially amongst commercial growers and in the countries of origin of the wild species, in order to protect those plants remaining in the wild.
5. To promote other, more specific, projects which may be brought to the attention of the Group through its wide membership and by others.

Ig. *Cephalanthera rubra* Red Helleborine

* 16th July Gloucestershire

Whosoever finds the fickle-flowering Red Helleborine in Britain in any state is in luck, and one is especially fortunate if both plants in a clump of two show at least some flowers – five each – though in this case they were not at their best. Both the petals and the sepals are of a pink that varies, according to the light, between carmine and rose, and are in no way comparable with the brick-red of *Epipactis atro-rubens*, the Dark Red Helleborine. Someone should change its name. The leaves, four or five in number, are spear-like of a mid-green colour and grow alternately from opposite sides of the stem which, though more or less erect can suffer from curvature of the spine. In profile the flowers suggest a thin tulip with sharply pointed petals.

These plants, which were on private land and were visited with permission, confirmed the conventional finding that this extremely rare orchid, rated on the continent of Europe as an endangered plant, favours a position close to the trunks of matured beech trees – that is on basic soil. Perhaps this is because moisture and humus collect there, or because competition from other plants is less intense. The experts seem silent on this matter. There was nevertheless a small oak seedling, about six inches high growing about the same distance away from one plant, and some ground-hugging brambles perilously near the other. One or two dead sticks littered the ground. The slope was moderately steep.

The orchid has a shallow rooting system (which is presumably why it is able to grow successfully under trees whose nourishment comes from a deeper level) but this makes it especially liable to invasion by fungi, particularly those of the Rhizoctonia family. During the autumn, when the fungus is at its most active, the orchid roots are nourished by sugary substances generated by the invaders, but, as temperatures rise with the spring, the plant becomes more active, and is able to produce a show of leaves – but less probably a flower. Whether it does so seems to depend on the relationship between the orchid and the fungus as well as on the amount of light reaching the plant. In Britain the Red Helleborine was, until recently, preserved in only two areas, the Cotswolds, where the soil is oolite limestone, and the Chilterns where it is chalk. In 1986, however, an entirely new, and fully authenticated site was found in a south-coast county. A colour slide of a plant on this site shows a robust orchid, with a profusion of blooms. It may have existed there without flowering for a considerable number of years, until, quite by chance the site was partially cleared.

It is therefore entirely possible that the Rosy Helleborine, to give it a more suitable name, exists in areas where it has escaped notice for many years, perhaps in fairly deep shade.

Where however the light, or even half-light, penetrates, the chances are better.

Searchers in Gloucestershire believe that they have found the leaves of the plant in more than one other locality but without flowers. Also there are records claimed outside the two best-known areas none of which until recently have been confirmed. Thus *The Flora of Sussex* by Lieut. Colonel A.H. Wolley-Dod records several specimens of this species were seen in the hands of a woodcutter near Poling (a village two miles north-east of Littlehampton) in 1921 by H.M. Edelstein, who is quite familiar with the species in Gloucestershire, and the author concluded that this identification might be relied upon. He added however that the plant was quite unkown to local botanists and has not been seen since, and so concluded that it was best excluded from the list of Sussex Flora for the time being! It has still not been found, but, as long as woods are cleared and trees felled, there may still be hope. To encourage the search for new discoveries is one of the objects of this book, and it does no harm to remember that the site in the Chilterns which at one time included more than 60 plants remained undiscovered until 1955, the site in the southern coastal county until 1986. To the shame of the local botanists the new site was discovered by a group of wandering ornithologists.

Ih. *Calamintha sylvatica* Wood Calamint

* 13th September Isle of Wight

Away from its yachting centres and holiday rendez-vous, the Isle of Wight is a network of narrow lanes, requiring close attention from the motorist, so that he hardly dares to look closely at the hedgerows and banks even when travelling in low gear. Nevertheless, while driving up a winding lane leading on to one of the central downs, the sharp-eyed observer might notice in one of the bays which allow one car to pass another, a very beautiful wayside plant. With luck this could be *Calamintha sylvatica Bromf.*, Wood Calamint. In a way it would have been simpler for us all, if it had been given a more distinctive name because there is another Calamint, a close relative with the name of *Calamintha ascendens*, which really looks somewhat like it except for the flowers which are paler and smaller and the leaves which are also smaller with fewer and less distinct serrations on the side. So it might have been sensible to have called one plant the Greater Calamint and the other the

Lesser Calamint. Perhaps there is another difference other than size to which attention might be drawn – which is the way the rarer plant proudly displays its blossoms – almost like those twin motor horns which were such a feature of vintage sports cars half a century ago.

But there it is. Perhaps *Calamintha sylvatica* does grow in woods in France, Germany, Spain and other parts of north Africa and Syria where it is to be found. In Britain it has just this one station by the roadside on the edge of a wood, though it might spread if established in a coppiced wood.

It will have been noticed that the plant is known as *Calamintha sylvatica Bromf.* The last word of the three is an abbreviation for the name of the man who first discovered it and correctly described it in a validly published paper with a note as to where it could be examined. Many plants carry at the end of their botanical name a mere capital (L.). This refers to the fact that Linnaeus named and described the plant and it is as much a compliment to him as a labour-saving device for the scribe that his name is cut short. W.A. Bromfield's name is not so well known, so he is entitled to a more specific identification even though his full name is not spelt out.

He was in fact a doctor who lived in the Isle of Wight. He discovered the plant on 29th August 1843 as 'growing profusely in woods on the west side of a small valley'. During the 1840s he was collecting material for a *Flora* in which he intended to cover both Hampshire and the Isle of Wight. But in 1850 while on a visit to the Middle East, he contracted a fever from which he died. Luckily by that time he had already completed his survey of the Isle of Wight and, six years after his death, this was published as *Flora Vectensis*, the only complete treatise dealing exclusively with the Isle of Wight.

Calamintha sylvatica is on the list of Scheduled Plants. And so it should be, for visitors to the site in the days before the Second World War remember it stretching for about half a mile in its happy valley, instead of, as in more recent times, for a comparatively few yards.

Ii. *Epipogium aphyllum* Ghost Orchid

* 20th August Oxfordshire

Here we have an extremely rare orchid that possesses no chlorophyll and consequently is unable of itself to convert elements from the air above and the earth beneath into stems or petals or even leaves. Its unconventional roots look like pieces of coral or something out of a jigsaw puzzle, and it depends for nourishment on matter produced by other plants, and served up to it by fungi. But its life-style is only one reason why the name Ghost Orchid (bestowed on it jointly by Rex Graham and David McClintock) is so especially appropriate. What struck the two botanists as they looked at the orchid, a few inches high, growing in a dark beech wood, was the way it so perfectly matched the litter of dead beech leaves beneath with their white veining and changing lights. And so it became the Ghost Orchid.

Up till that time it had been inappropriately described as the Spurred Coralroot – thus confusing it with another species, Coralroot Orchid which admittedly has no spur, but differs in many other more important respects, in particular the presence of some green chlorophyll and the fact that the lip is at the base of the flower. In the case of the Ghost Orchid, the lip, of translucent white – another ghostly characteristic – and marked with raised dots of violet, is at the top of the flower and presents a floppy outline akin to that of an old-fashioned sunbonnet. Behind it is a rounded spur, standing upright like a half-inflated toy balloon. The remaining two petals and three sepals droop downwards.

The plant has had an almost legendary history. It was unknown in this country until 1854 when it was discovered by Mrs W. Anderton Smith by the Safey Brook near Tedstone Delamere, midway between Worcester and Leominster. Unfortunately, the plant was dug up and transplanted to a garden, where, as might have been expected, it failed to survive. The species has not been recorded again from this locality. However, in 1876 another plant was found at Ringwood Chase, near Ludlow in Shropshire. A further plant was discovered in the same wood two years later and a third in a different part of the same wood in 1892. A single plant was recorded in July 1910 in the Wye Valley. No further plants have since come to light in this area though it may well have been overlooked.

All these plants, it will have been noticed, were in the western part of the country where the rainfall is usually higher than average, and it seems likely that the plant, having only a few hairs on the rhizome, finds difficulty in absorbing the water which it needs if it is to send up a flowering shoot, and can do so only if it receives a more or less regular supply of water, although it can store some liquid in a swelling at the base of the stem.

Flowering tends to occur most often in seasons during which there has been a heavy fall of rain in the spring. It can occur at almost any time from June onwards to early October. The flowers are said to carry the scent of bananas though having no finely tuned sense of smell, I myself could not detect this. They are also said to be pollinated by wild bees – although it cannot be often that sufficient flowers are in evidence for cross-pollination between different plants to take place.

However, the plant is capable of reproduction by other means. Thin reed-like runners are sent out from the rhizome and from these at intervals new plants are formed. These, in turn, develop coral-like rhizomes which, nourished like the parent with food provided by fungi from decaying humus, gradually increase in size and in time become capable of sending up a flower shoot, either while still attached to the runner or as a separate individual plant. This process may stretch over a period of years while the rhizome grows to the point where a spike can be sent up. It therefore follows that the orchid may disappear for years and reappear in an unlooked-for site – like a ghost.

Indeed, there is a strong suspicion that at one site in the Henley area the runners have passed beneath the surface of a narrow road which separates one part of a beech wood from the other. But we have got ahead of the story.

The next notable event in the history of the Ghost Orchid occurred in 1924 when Mrs Vera Paul, then a schoolgirl, found a plant in an area near Henley in Oxfordshire, and, seven years later, another plant in a different wood in the same area. This specimen was 23 cms (7 ins) tall and bore three flowers. Then, for twenty years, no further plants were seen. In 1953 Rex Graham who had been searching for the plant for two decades was in a Buckinghamshire wood about ten miles from the Oxfordshire site. He was lighting his pipe and as he looked at the bowl, he saw beyond it a plant of *Epipogium* growing among the beech leaves. It was one of 24, twice as many as had been counted during the previous century. More recently the reports have been less encouraging with 1984 and 1985 being reportedly blank years.

Ij. *Lloydia serotina* Snowdon Lily, 'Lloydia'

* 3rd June Gwynedd

Lili'r Wyddfa, the Snowdon Lily, as the Welsh call it, once flowered on a dozen peaks in Caernarvonshire. Now there are only five colonies – 'somewhere in Snowdonia', despite the hazards to life and limb which protect it from the attentions of those who want to approach it too closely.

All the agonising frustrations of the botanist looking for Lloydia and the anxieties of the climber on Snowdon are chronicled in the book *Fiery Particles* in which some of C.E. Montague's sketches and essays were published in 1923. This particular short story begins with Montague, a rock climber by inclination, sitting by himself at the inn at Llyn Ogwen, a convalescent at the time and under doctor's orders to avoid all violent exercise. His mind however was filled with mutinous intentions for, after all, from Ogwen it was but a half-hour's walk to the foot of a climb which clearly in Montague's opinion would not involve any disobedience of the medical advice he had received. He did not, however, feel inclined to go climbing alone; that, as the initiated are well aware, is against the conventions of the sport as well as the dictates of common sense. At least that is what coroners say about lone mountaineers who return, inert, on a stretcher. Nevertheless, Montague was somehow able to convince himself that an exception must be made in his case and ordered sandwiches for eight o'clock the following morning. And then . . . soon after he had resolved this moral issue in his own favour, a stranger arrived. He was a self-confessed botanist armed with a cylindrical green vasculum – a good fellow, apparently, tho' with only one good leg. His name, he said, was Darwin and though he had been christened Charles, he had felt obliged, on coming of age, to change it by deed-poll to Thomas. A fellow with modesty of this kind was, Montague felt, the best possible companion for a climb. Moreover Darwin was keen to make an ascent, for he was anxious above all to find and preserve – in his herbarium – a specimen of *Lloydia serotina*.

Anyone sharing those ambitions should, of course, forget all about the vasculum and the herbarium, at least as far as *Lloydia serotina* is concerned, for the penalties for disturbing such a prize are now severe. But in mitigation, we accept that Montague's story was written, and Darwin's misdeed committed, more than sixty summers ago. The lessons of the climb are, however, much the same today. Montague, who was no botanist, but had some idea of where good plants might be expected to grow, suggested that they should make for Cwm Idwal – not for the much-feared Twyll Du known as the Devil's Kitchen which was wet, rotten, lethal, and barren, but for the conspicuous hollow in the cliffs nearby to the right of the Kitchen known (though seldom on maps) as the Hanging Garden Gully. This, Montague said, was the kind of place where town children could spend a happy day in the country and would be just the place for Lloydia.

The two conspirators went at once to the lady of the inn, and found that she had no mere clothes'-line but a length of 120 ft of Alpine quality climbing rope purposely left behind by a mountaineer for his own use should he return, or in his absence for the use of others.

The next morning they climbed, and Darwin was made fast to one end of the rope and Montague hitched himself at the halfway point of the rope, with 60 ft falling loose. We must leave Montague himself to describe how he inched up the crack in the face of the crag, encouraging Darwin to copy his movements as far as possible; how he jambed his half of the rope in two places and threw the loose half down to Darwin, telling him to make himself fast to it and then release himself from the original rope; how Darwin neglected to do so, and then against instructions, clambered across a smooth plate of rock to seize a piece of Lloydia, returning with it in his teeth; how he reached an upper grassy slope which was speckled with Lloydia, how in his ecstasy of running up the greasy slope, he jambed the lower end of the rope so that it was impossible for Montague to free it without descending again meanwhile leaving the botanist in perilous isolation so the rope had temporarily to be left behind as a public token of mountaineering ineptitude, Montague then pressed on towards the top, ropeless, driving Darwin before him, happy nevertheless in the thought that when he came back later to retrieve the rope, he would be climbing alone.

The Hanging Gardens, for there are others besides the one at Cwm Idwal, are to be seen in several parts of Snowdonia. They are relics left to us by the glaciers of the Ice Age. These glaciers were once fed by small tributary streams, each trickling down its own private valley till its waters froze at the head of the glacier. In time, the glaciers, with the ever-increasing weight of their ice, gouged away the ground beneath them with each inch of movements, so that finally, the head of the glacier pressed deep into the body of the mountain, sinking down into it and breaking away from the slope above it and thus shearing off the lower ends of the small tributary valleys, leaving them, as it were, hanging.

Eventually, the glacier, dragging stones and rocks along with it scoured out its own track – straight as a pig's trough downhill with never a twist or turn nor even the wrinkles and folds that a conventional valley has along its flanks. The bed of each glacier valley is U-shaped, with more likely a lake than a stream at the bottom, and the top of the valley, at the point where the head of the glacier parted company from the hanging valleys, we see a curve, as rounded as the dress circle of the Scala, for that is the form in which the ice of the glacier finally melted.

The great Charles Darwin was astonished that he had not drawn the correct inference from what he had seen in 1831 when he paid a visit to Cwm Idwal. The scored rocks, the boulders displaced and perched incongruously around the sides of the valley, and the formations at either end were so conspicuous, he said, that a house burnt down by fire did not tell its story more plainly than that valley. 'If it had still been filled by a glacier the phenomena would have been less clear than they now are.' So the Hanging Gardens are part of a relatively unchanged original landscape and thus the haunt of rare plants of the past. And the visitor to Snowdon should lift up his eyes to the heavens in the hope of finding them.

Cwm Idwal is, of course, not strictly speaking a part of Snowdon itself but is separated from it by the road leading through the pass which flanks the great slate quarries of Dinorwic on the way down to the village of Llanberis. But for earlier botanists Cwm Idwal seemed to be the area where the best plants grew. Thomas Johnson, whose journeyings in Kent are described elsewhere in this book, paid a visit to the Snowdon area in 1639 and referred to a feature described by the locals as 'the home of the devil', though he seems to have thought that it referred to a marsh. In any case he failed to notice Lloydia and it was left to Edward Llhwyd (1660–1709) to record it for the first time. (He preferred this spelling though Lloyd was the style of his father).

Lloyd was an Oxford-trained scholar, descended from an old crusading family. He became Keeper of the Ashmolean Museum in Oxford at the early age of 22. The portrait of Lloyd in the possession of the Ashmolean Museum shows him in academic robes, his face, smooth, plump and benevolent framed in generously curled wig. No one would have suspected him of wearing climbing-boots, but he spent the years 1695–1699 on a Grand Tour of Wales. During this he sent lists of some of his discoveries to the great naturalist John Ray, who included 'Lloydia' in the second edition of his work which bore the impressive title *Joannis Raii Synopsis Methodica Stirpium Britannicarum, Tum Indigenis, tum in agris cultis, locus suis dispositis, Additis Generum, Characteristicis, Specierum discriptionibus, & Virium Epitome. Edition secunda.* Ray's work was published in 1696, so it is evident that Lloyd had come across the plant no later and probably before that year. This edition was of such value to botanists that, after Ray's death in 1705, there were ever-increasing demands for someone else to undertake a new edition, incorporating details of new plant discoveries which had materialised in the meantime, and which students endeavoured with varying success to write into the margins of their existing copies. William Sherard, who had studied botany under Tournefort in Paris and explored much of Europe and Asia minor during a fourteen-year assignment in Smyrna, returned to England in 1717 and persuaded John Jacob Dillenius to come to England to help with a revised version of both Bauhin's vast botanical work *Pinax* and of the much shorter *Synopsis Methodica*. This appeared in 1724, and, in it, the Snowdon Lily appeared, still credited to Llhwyd as *Bulbosa Alpina juncifolia* – the rush-leaved bulbous Alpine plant.

But as to the full name of Lloydia there seems to have been a misunderstanding somewhere; because, when the great Linnaeus came to write *Flora Anglica* (1754), which was his first attempt to reconcile the names used in Ray's *Synopsis* with his own new binomial system, *Lloydia* had appeared in Linnaeus' principal work *Species Plantarum* as *Bulbocodium serotinum*, meaning late-flowering Bulbocodium. The name chosen for Lloydia in the *Flora Anglica* is even more inapposite: *Bulbocodium autumnale*. The generic name was subsequently changed several times but became Lloydia on the instigation of Richard Salisbury (1761–1829), one time Honorary Secretary to the Horticultural Society of London. The adjective *serotina* has, as we have seen, been retained right up to the present day. Linnaeus chose it in order to contrast it with another Bulbocodium which also carried lanceolate leaves and flowered in Spain in spring.

Dr Richard Richardson, a Yorkshire botanist, who had been shown the plant by Lloyd, described it in a note – also included in the third edition of Ray – as 'intus albo extus squalide rubente'. Gerard, the famous seventeenth century herbalist, would probably have described the red on the outside of the petals as 'overworn'. In the bud stage, the needle-like sepals themselves appear a dingy beetroot red, and the colour is suffused into the veining of the petals nearest to the sepals. However in the later stages of blooming the colour fades into a greenish yellow which tinges the inside of the corolla.

The shape of the petals also appears to change as the flowers mature. In bud, the flowers are the shape of a narrow tulip; in full flower the bloom is chalice-like with the petals still slightly overlapping. But in the later stages the petals lengthen and spread more widely to give a star-like appearance to the flower. The needle-like leaves – usually two – sprout upwards, before curving outwards and then downwards towards the ground, reminding one of the arthropod legs which earned the plant its alternative name of Spiderwort.

The first ten days of June are the best times for the Snowdon Lily but it is a pity to spend too much time searching for it when there are so many

Ib.

Ic.

△**Ie.** ▽**Ig.**

If.

Ib. *Helianthemum apenninum* (ht 20 cm)
Ic. *Oxytropis campestris* (ht 25 cm)
Ie. *Ophrys holoserica (fuciflora)* (ht 15 cm)
If. *Cypripedium calceolus* (ht 20 cm)
Ig. *Cephalanthera rubra* (ht 30 cm)

Ih.

Ii.

Ik.

△Ij. ▽Il.

Ih. *Calamintha sylvatica* (ht 45 cm)
Ii. *Epipogium aphyllum* (ht 15 cm)
Ij. *Lloydia serotina* (ht 15 cm)
Ik. *Lychnis viscaria* (ht 60 cm)
Il. *Lonicera xylosteum* (ht 150 cm)

other worthwhile plants thrusting their attention on the climber who follows the herd up to the summit. On the way, or a little off it, are the pink pads of Moss Campion, *Silene acaulis*; Roseroot, like house-leeks with yellow pincushions of flowers crowning its glaucous-green leaves; *Cardaminopsis petraea*, Northern Rock-cress; *Saxifraga stellaris* Starry Saxifrage twinkling from some little niche it has seized for itself; *Thalictrum alpinum*, Alpine Meadow-Rue; *Minuarta verna*, Spring Sandwort, and many others.

The localities for Lloydia given in the third edition of Ray's *Synopsis* include the Glyder (the mountain behind the Devil's Kitchen), but no expert has been able to identify exactly which of the Glyder rocks was meant by 'Trigvylchau'! A number of localities on Snowdon itself were mentioned too including Clogwyn-du'r Arddu ('the dark precipice of the Arddu'), a rock-face so sheer and gaunt that it can be seen by the naked eye from as far away as Anglesey. It is still the haunt of alpinists and commandos.

My own experience, limited to Snowdon itself, is that botanists who are prepared to use their eyes and a pair of binoculars will find plenty of Snowdon Lilies without leaving one particular mountain track and, if they carry a suitable telephoto lens, can take all the photographs they need without putting any plants at risk.

Ik. *Lychnis viscaria* Sticky Catchfly

* 21st May Powys

At a casual glance this plant has a number of lookalikes, Bouncing Bet the Soapwort for instance, or garden Phlox, or even the very common Red Campion. But its distinction lies partly in its rarity and partly in its stem, which, instead of the usual green shows a stretch of dark purple or near-black beneath each pair of leaves. This is the sticky patch which gives the plant its name – and the purpose is assumed to be to prevent ants from climbing up to steal the nectar which is reserved for butterflies and long-tongued bees. The flowers are red with a tinge of mauve, and the blooms spring from the axillae of the leaves, in opposite pairs, giving the impression of a whorl, but not a full dress one as the flowers tend to open in sequence rather than all together. Sticky Catchfly is usually to be found growing on cliffs of igneous rock, that is on rock which at an earlier stage was molten. It is not seen in England, but there are two stations in Wales. Of these the better known one is on Stanner Rocks in what was

Radnorshire, and is now Powys, a few miles to the west of Kington. On the map Stanner Rocks appears to be set in the middle of rolling countryside and is not part of a mountain range. But it is a steep little climb and one best tackled with the advice and consent of the local office of the Nature Conservancy Council, which, at the time of writing operates from the Llysdinam Field Centre, Newbridge on Wye, Powys LD1 0NB.

The second Welsh centre is at Breidden Hill – pronounced as in Dryden – near Montgomery, and also in Powys. Here the plant was not recorded until 1890, sixteen years after it was known at Stanner Rocks. At Breidden however the plants have suffered from local quarrying and have had to be taken up and replanted, let us hope on another igneous rock.

There are also sites in Kircudbrightshire (now part of Dumfries and Galloway), Selkirkshire and Roxburghshire (both incorporated in the 'Borders'), in Perthshire and in Midlothian on the crags of Arthur's Seat overlooking the capital.

11. *Lonicera xylosteum* Fly Honeysuckle

* 24th May West Sussex

No one would mistake this for an ordinary honeysuckle. It is no climber but an untidy bushy shrub with greyish wood and grey-green leaves. Moreover the flowers, instead of developing at the end of the branches, are strung out along them in pairs, the buds standing upright, close together, like cream-coloured Indian clubs. The berries are globe-shaped and red.

This species must like chalk because the only site in which it grows wild in Britain is about halfway up a Sussex Down in a copse more or less in shade, for it is a north slope. The half-dozen or so shrubs were never part of a hedge and are miles away from any house, so it is classed as a Native plant (i.e. not known to have been introduced by human agency). It could, one supposes, have been bird-sown from some garden but in that case why should it turn up in one place only?

It was first mentioned in 1801 by William Borrer, the Sussex botanist (1781–1862) and confirmed a century later by J.E. Lousley, famous and popular field botanist and author of *Wild Flowers of the Chalk and Limestone*, who found it growing more freely in two or three other places nearby than on the original site. Local sowing by birds could account for this development, if not for the original introduction.

Fly Honeysuckle does occur elsewhere outside gardens in England, Wales, Scotland and Ireland but is presumed to have been introduced there. 'It is very common in our shrubberies,' according to the excellent *Handbook of The British Flora* by George Bentham 'for the use of beginners and amateurs', which also tells us that the genus Lonicera was named after the German botanist, Adam Lonitzer (1528–1586). However, as a wild plant this species is regarded as being at considerable risk, partly because of the restricted number of sites, and because they are not within a reserve.

Some books have this shrub flowering in May and June, but I have found it in berry as early as the first week of June, and early or mid-May seems the best time to see the characteristic buds as well as the open tubular flowers. My picture was taken in a late season.

— II —

The Most Beautiful

* (See following p. 82 for illustrations)

IIa. *Draba aizoides* Yellow Whitlow Grass

* 21st March Glamorgan

This neat, little cushion-plant with bright yellow four-petalled flowers, above a rosette of narrow dark green leaves, looks as though it had been taken straight out of someone's rock garden. All too frequently however, it is dug up and taken back there where it probably languishes. Only limestone rocks and walls suit it and, apparently, the comparatively mild climate of the Gower Peninsula in South Wales. The ruined castle of Penard, or Pennard, lying between the golf course and the river running into Threecliff Bay, is the best-known site, and the plants, high up on the castle walls, are probably safe for life. But there are many other localities on the south coast of the Gower west of Penard right along to the Worm's Head where *D. aizoides* flourishes on cliffs and on the turfs between the rocks.

There are two nature reserves on Oxwich Bay, and, although a permit is required if it is intended to leave the dunes or the paths through the woodlands, there is a car park and an information centre which will probably help to save the visitor's time.

Draba aizoides is already on show by the third week of March, and a prompt visit finds the flowers at their best; leave it for too long and the blooms will have faded from sharp yellow with a hint of green to the bland colour of summer butter.

It would, therefore, be advisable when planning a trip, to write or telephone in advance to the Nature Conservation Council's Headquarters for Wales at 43, The Parade, Roath, Cardiff, for which the current telephone number is Cardiff (0222) 485111.

'Doubtfully native' is the verdict of Clapham, Tutin and Warburg's authoritative *Flora of the British Isles*. But the Welsh have apparently accepted it as one of themselves and have found a name for it: Llysiau

Melyn y Bystwn. The *British Red Data Book* accepts it as presumably native and certainly there is no direct evidence of its introduction by human agency.

The reputation of the *Draba* genus for curing whitlows in the finger-nails is of long standing and, in Gerard's day, both *Draba* and *Saxifraga tridactylites*, Rue-leaved Saxifrage, were grouped together in the genus *Paronychia* from the Greek word for whitlow. Gerard says that one of them – he does not specify which – 'groweth plentifully upon the bricke wall in Chancerie Lane belonging to the Earle of Southampton'. Linnaeus however appears to have doubted whether a cure for whitlows was guaranteed, and considered that it was more relevant to call the genus to which our plant belongs *Draba* from the Greek word for acrid. Gerard certainly found that his Whitlow-Grasse tasted 'somewhat sharp'.

IIb. *Dianthus gratianopolitanus* Cheddar Pink

* 11th July Somerset

'It is very different from the Maiden Pink, and more truly answering the name, having but one single flower on top of the Stalk,' wrote John Ray in his *Synopsis Methodica Stirpium Britannicarum* (1724), adding that the pink had been found in 'Chidderocks in Somersetshire' by a Mr Brewer.

And it is still there two and a half centuries later, with good hopes of remaining untouched. One reason is that many of the Cheddar Pinks above the Cheddar Gorge cannot be reached by the casual visitor unless he or she is equipped with climbing gear. Secondly, the site is frequented by numerous visitors and the risk of being seen digging up a plant and thus incurring a fine of £1,000 is very real. Finally it is possible to buy from a reputable nursery plants which, one hopes, have been raised from cuttings taken from other specimens already grown there.

And what a wonderful feeling it is to be high up above the souvenir shops of Cheddar walking up the gentle turfy slope past thorn bushes and a forest of trees, until finally one is out on the open downs. Near the sheer cliff edge, amid the grass, on the rocks, grows this wonderful display of pinks, on show from the first week in July onwards.

As it happens there are also several interesting plants on the way up, for example, *Thalictrum minus*, Lesser Meadow Rue, with its pale yellow tassels dancing above the fernlike leaves in the breeze. . . . *Prunella laciniata*, the white Self-Heal which is no mere albino but a separate species distinguished from the ordinary purple Self-Heal by its cut leaves. But as for the Pink itself, few plants can offer such a distillation of the colour rose. (The flower should, however, if possible be photographed *contre-jour*, that is into the sun so that the light reaches the camera through the petals and not as a looking-glass reflection from them which is the case when the flower is photographed from above.)

The flowers are about 2.5 cm (1 in) in diameter and deeply scented and are attractive to day-flying moths such as the Hummingbird Hawkmoth, as well as to butterflies. This is a perennial with a woody stock and long non-flowering shoots growing along the ground, so that special care is needed to avoid damage to these when approaching a plant. The adjective *gratianopolitanus* means pertaining to the city of Grenoble, and for all I know, the Cheddar Pink grows there very well too.

IIc. *Dryas octopetala* Mountain Avens

* 3rd June County Durham

Here is a ground-hugging member of the Rose family, an undershrub which, despite its small size, develops a woody base and is tough enough to last out the winter with its buds unprotected. This readiness to make the most of a short summer has helped it to spread not only worldwide across the Arctic in Europe, Asia and America, but also in the higher mountain ranges of Europe including the Pyrenees, the Apennines, the Balkans and even the Rocky Mountains.

Dryas can grow on bare unsheltered rock provided that, over the years, sunshine and frost have opened a crack for it, and the weather has crumbled away a few fragments to nourish its roots. There is but one proviso: the rock must be basic – that is containing mineral salts; ordinary granite is not acceptable. But this was no disadvantage through the Ice Age when the glaciers during their descent from the mountains brought down large quantities of chalk and smeared it across large stretches of the countryside. So we find the white flowers of *Dryas* – 2.5–5 cm (1–2 ins) – across on ledges, screes and sometimes even on grassy banks. The *Dryas* stem has many twists and turns and numerous branches so that an almost impenetrable patch is eventually formed.

The *Dryas* has been with us in Britain since before the end of the final Ice Age, since when it has been retreating gradually northwards before our warmer summers and the advance of the forest trees. In West Yorkshire it has sought refuge on the limestone of Pen-y-Ghent and Ingleborough, and in Co. Durham on the high fells above the

Tees, but in the very north of Scotland near Bettyhill it approaches sea-level.

One would naturally suppose that the name *Dryas* has something to do with Dryads – the Greek nymphs who were thought to live among the trees – but then this would seem an unsuitable name for a plant of the wide open spaces.

Another possible and more plausible explanation is to be sought in the fact that *dryas* is also the Greek word for an oak and that the leaf of Mountain Avens is oblong and scalloped just like an oak leaf. The flowers – not always eight-petalled – stand on stalks above the closely packed leaves and each is arranged like the reflector of a heat-lamp and like a good reflector it follows the sun on its course, giving out light and perhaps some heat to encourage insect visitors. In areas where it is unlikely to get any, the flowers are capable of self-pollination.

Alpine gardeners cherish the Mountain Avens not only for its neat habit and dark green leaves, shiny above and white and silky beneath – but also for its fruits which, like those of the Pasque Flower, carry silky plumes.

The Mountain Avens flowers between May and July the date depending to some extent on the altitude and local conditions. The one shown in this book was photographed in Upper Teesdale.

IId. *Epipactis atrorubens* Dark Red Helleborine

This orchid is apparently not considered to be officially at risk as it has assuredly been recorded in recent years from fifteen or more 10 km square sites. Yet it is a rare plant, and the sites so scattered and the plants on each site so scanty that one must have misgivings about its ultimate survival.

One finds it perhaps in an ash wood in the Yorkshire Dales, and I have seen it in a cleft of the rocks by the very edge of the sea near Cape Wrath. In Wales, it has been recorded in Breconshire, Caernarvonshire, Flintshire and Denbighshire – making nine Welsh sites in all.

It is also to be seen on the geological formation known as limestone pavements which are to be found in the clearings around Morecambe Bay, around Ingleborough, Malham, Shap Fell, in Wales and the West of Scotland. These 'pavements' are relics of the Ice Age: they are cracks left by the glaciers as they drove across the soft limestone crushing it and grinding it down to give an almost smooth surface. Peculiar boulders trapped in the glacier and carried for hundreds of miles are strewn around, left in strange positions where they rested as the ice melted. But, limestone being what it is, in time, cracks began to appear in the surface of the pavement and the frost and the wind and the rain, which even under natural conditions contains some carbonic acid, has dissolved some of the limestone cutting out deep cracks known as 'grikes' between the smooth 'clints' so that a stretch of white limestone pavement has the strange beauty of a landscape on the moon.

In the shallower grikes many beautiful plants such as the so-called Bloody Cranesbill (carmen would, I think, be a better match), the orange Alpine Cinquefoil, various saxifrages grow in profusion in a natural rock garden. However, a caution is needed. On one beautiful stretch of limestone pavement near Chapel-le-Dale there are all the above flowers and sheets of Lily of the Valley as well. The drawback there is that some of the grikes have become turfed over, but if you step on the turf you wish you hadn't. Your foot goes down into the void, you lose a lot of skin, and extracting it may prove difficult.

As it happens, the Dark Red Helleborine is not to be found, as far as I can tell, on this Chapel-le-Dale pavement but there is another very good stretch of pavement in West Lancashire where it does flourish – on the Nature Conservancy Council's Gait Barrows Reserve – for which you will need a permit. Limestone pavements have unfortunately been exploited commercially, for building walls, gateposts, and more recently for rockeries, and a thousand tons of irreplaceable clints were sent from Ingleborough for the Festival of Britain in 1951 as part of an exhibit entitled 'The Origin of the Land'. In some cases explosives have been used with powerful tractors to take away what can never be restored. However Gait Barrows was rescued just in time.

The flowers of the Dark Red Helleborine are of a deep wine colour and make a fine show against grey or white rocks, and illustrate well the way in which the genus *Epipactus* differs from *Cephalanthera*, though both are known as Helleborines.

The flowers of *Epipactis* are stalked, hanging and open. Those of the *Cephalanthera* genus are without stalks and reluctant to show their wares.

IIe. *Gentiana verna* Spring Gentian

* 19th May Upper Teesdale

Those disinclined to make the journey to Clare or Galway can see this sapphire jewel-flower on the sugar limestone of Upper Teesdale where it is locally plentiful. The main colony is within the Nature Conservancy Council Reserve where a fine display of flowers can be seen without leaving a hard surface route. Yet, although the road in question is almost permanently under the public gaze of visitors, and is patrolled by wardens, anxieties remain about the risks of plants being collected (even though slightly improved varieties of this species are on sale from at least one nurseryman).

So, although the map reference for the reserve is to be found in the *Guide to Britain's Nature Reserves*, it is as well to get in touch first with the Durham County Conservation Trust's Visitors' Centre for Upper Teesdale, which is housed from Easter onwards in a converted chapel in Bowlees, four miles up Teesdale from Middleton. There is no need, however, to confine one's search to the reserve; indeed, I was shown some fine Spring Gentian plants in a meadow right opposite the small hotel near the reserve in which I happened to be staying. There are also colonies in Yorkshire and Cumbria. The third week in May is a convenient time for a visit to Teesdale and interesting plants such as *Primula farinosa*, Birds Eye Primrose; *Viola lutea*, Mountain Pansy – usually purple in this area and *Antennaria dioica*, Mountain Everlasting, among others, will also be on show at that time.

IIf. *Gladiolus illyricus* Wild Gladiolus

* 12th July Hampshire

Gladiolus – meaning little sword – was the adjective used by the Romans to describe the leaves of this plant. 'Gladdon', a corruption of Gladiolus, was taken from the Romans by the Saxons, and is still used among countrymen to refer to a far more common relative with sword-like leaves, *Iris foetidissima*, which is called the Stinking Gladdon to distinguish it from the Gladiolus.

This fine plant with crimson flowers which grows amid the bracken on heathland is now found only in Hampshire, though it was formerly recorded in Dorset and Devon as well as in the Isle of Wight, where it grew in the Lake, Blackpan and Apse Heath areas, all three near Sandown as late as 1897. Some forty sites were known at the time when the *Red Data Book* (1983) was prepared, but the plant is usually well concealed beneath bracken and many other stations may well have been overlooked by searchers not flexible enough to adopt an all-fours approach.

The plant propagates itself mainly from offsets from the corm in the same way as the garden tulip, though seeds are no doubt produced if insects can find their way through the underworld beneath the canopy of bracken. The flowers are what the botanists call 'secund'; that is they all face in more or less the same direction, although each is lightly turned away from its neighbour. The six petals are loosely arranged with the upper central petal protruding in front of the others. The three lower petals are slightly reflexed and the two lateral petals are almost hidden from view. The flowers appear in the first fortnight in July, the best-known sites being in the New Forest where the plants are at risk if they are close to one of the many picnic sites. Luckily *Gladiolus byzantinus*, a similar species, but with slightly larger and differently formed flowers, is easily obtainable from nurserymen.

IIg. *Helleborus viridis* Green Hellebore

* 16th March West Sussex

The silence and half-light of a beechwood on the chalk, with the tree trunks stretching skywards like pillars in a cathedral, shielding the plants beneath from the winds – that is the right scene for the Green Hellebore. The beechwoods above Gilbert White's Selborne and the hanging woods of the South Downs facing Petworth are two good settings, with one colony of some forty plants in the Sussex station.

It has been recorded in 42 of the 10 km x 10 km squares in Wales and this alone would be enough to exclude it from the *Red Data Book*; nevertheless, it is a plant at risk and a sudden decision to fell a beechwood which may have been standing for half a century could wipe out a site almost overnight. As a native plant it is, in any case, limited to the countryside south of Hadrian's Wall. Often its leaves grow up amidst the green drugget of Dog's Mercury so that, at an early stage it is difficult to distinguish them. But, when fully grown, we may expect to see a plant fully 40 cm (18 ins) tall with green flowers two or so to a plant up to two inches across.

In the south of England the flowers are probably at their best in mid-March but, on occasions, one or two have been seen as early as Boxing Day.

The Green Hellebore was one of the plants to appear in that revolutionary work *Herbarum Vivae Eicones* – Living Images of Plants, published in Strasbourg in 1530. It was written by Otto Brunfels of the same city but the illustrations by Hans Weiditz were among the first to show drawings made from real plants with all their imperfections in place of the conventional mediaeval devices as remote from the truth as the diamonds in a pack of playing cards.

The flower of Weiditz's Hellebore, however, is in perfect condition with the sepals open like the dish of a shallow saucer – these flowers have no petals – the stamens not, as yet disarranged, and the bloom not yet drooping as it does in the later stages of flowering. The artist's viewpoint is centred on the lower third of the plant so that the stalks and leaves thrust upwards as if still in growth.

The leaves are serrate – saw-edged – and pointed at the tips which perhaps accounts for the countryman's name for this plant – Bearsfoot – the serrations giving just that suggestion of the fur round the bear's foot with sharp claws beneath. One must hope that the gardeners of the future will continue to prefer the insipid Christmas Rose and other continental Hellebores to this beautiful wild plant. If they do, it may be because its foliage unlike that of other Hellebore species dies down and disappears in the winter, making it necessary to mark its position in the herbaceous border. This feature is one of those that help to distinguish it from its near cousin *Helleborus foetidus*, the Stinking Hellebore – equally rare and green-flowered. The flowers of this latter plant are bell-shaped, with reddish-purple rings round the tips of the sepals and they grow in thick clusters. Both plants have a burning taste and are poisonous, but were nevertheless extensively used as a purgative and emetic by ancient apothecaries.

IIh. *Melittis melissophyllum* Bastard Balm

* 11th June Hampshire

Far more handsome than the real Balm, this plant really deserves a more attractive vernacular name, especially since it *is* a native plant, whereas the puny-flowered 'true balm' is an introduced interloper. Balm-leaved Archangel, the name mentioned by John Ray, might be better –(Archangel referring to that Yellow Dead-Nettle which makes such a show in woods after they have been coppiced). Most writers comment on the 'strong smell' of *M. melissophyllum* and the Rev. C.A. Johns in *Flowers of the Field* goes so far as to say that the smell of the leaves when fresh is 'offensive' – adding however that 'in drying they acquire the flavour of new-mown hay or Woodruff' (a sweet-smelling Bedstraw).

The plant grows to about one foot, with leaves of dark green, egg-shaped and with scalloped edges. The flowers, growing in whorls of two to six, spring from the axillae of the leaves, are up to 2.5 cm (1 in) long; they are white and heavily blotched on the lip with colours varying from pale pink to deep crimson.

M. melissophyllum favours the banks and edges of woods which provide it with shelter as well as with a background which ideally sets off the conspicuous flowers. *M. melissophyllum* prefers the moist atmosphere of the West Country, appearing most frequently in Devon and Cornwall (where the small country lanes are nicely banked) and also in the westerly counties of Wales. It has been reported intermittently from Sussex. The plant in the photograph was one belonging to a small colony in the New Forest. Bastard Balm is another of the plants which, though rare, is sufficiently scattered to have appeared in fifteen or more of the 10 km squares into which the country was divided when the last 'census' was taken, and it had not at the time of writing been included in the Schedule 8 list of plants requiring special legal protection. Let us hope that it soon will be, even though the smell of the leaves may afford it some protection from casual marauders.

lli. *Orchis militaris* Military Orchid

* 5th June Suffolk

Perhaps the name 'Military Orchid' is unsuited to the unregimented and dishevelled appearance of this orchid when in flower. However, each individual blossom has above it a 'helmet' of three sepals growing together – ash-coloured on the outside, with a glint of some metal, which, if not steel, might be lead, or even the dull side of the foil in which a chocolate soldier is wrapped. A close inspection of the lining of the helmet shows that it is striped with precision with reddish-purple lines – as in the underside of an awning. The two upper petals are shrouded from sight beneath the helmet, but the labellum beneath has the kind of swashbuckling style of which any military man might be proud.

It has right and left arms branching out from it at the top, white, spotted reddish purple, often with dark red 'cuffs' at the end of each sleeve. The centre of the labellum is also white, spotted, in two lines like rows of buttons on a tunic. At the bottom, the labellum divides into thick 'legs' with, between them, a sharp tooth, which could, perhaps represent the soldier's scabbard.

This orchid was once not uncommon on the chalk range of the Chilterns from Streatley in Berkshire, north and east to Oxfordshire, Buckinghamshire and Hertfordshire. But it gradually decreased in numbers and by the mid-1920s was believed to be extinct. Then in 1947 while picnicking in Buckinghamshire, J.E. Lousley discovered a colony of thirty-nine plants, eighteen of which were flowering. The site was in a small wood, and Lousley noticed that the trees in one part had been cut down, probably soon after the beginning of the Second World War in 1939: he concluded that the extra light had caused the seed to germinate (or possibly had stimulated the rhizomes into action).

Certainly, plants that I have seen abroad, near Lake Annecy, have been fine specimens growing in the open on a grassy south-facing bank of a small lane winding amid orchards and dairy farms. Furthermore there have been occasions when the Buckinghamshire plants have spread from the area of scrub and yew trees in the midst of their wood on to the side of the road. Over-shading such as develops with time in most woods is clearly a risk. This makes it so strange that the only other site known in Britain at the time of writing is not only miles away from the Chilterns, but in Suffolk, in a deep shaded chalk-pit surrounded by Forestry Commission land.

The plant is believed to show leaves about four years after seeding and a further four years is likely to pass before flowers appear. During this last period, the seedlings are vulnerable to trampling and in both Buckinghamshire and Suffolk, measures have been taken to keep out unauthorised visitors. On one day in the year, however, the Suffolk site, near Mildenhall which is under the protection of the Suffolk Trust for Nature Conservation, is open to visitors, who are permitted to enter the reserve and admire the plants from the safety of a board-walk. A small contribution is expected from those who take photographs. This site was discovered in 1955 during a routine survey made for incorporation in the *Atlas of the British Flora* which was being prepared by members of the Botanical Society of the British Isles. But for this piece of good fortune the site would almost certainly have become completely overgrown and the orchids would have vanished for good. (Although the Suffolk site is well protected by a rigid unscalable chainwire fence and solid padlocks, unwanted visitors still get in there in the shape of the winged seeds of the sycamore trees not far away, and, as J.E. Lousley noted when he first discovered the Buckinghamshire colony, rabbits evidently delight in biting off the flowering stems). It is always possible that new sites for the Military orchid may eventually be discovered. But a general increase in the population seems unlikely. The figures for seed produced and germinated are not encouraging and it may well be that our cold springs and tepid summers do not suit the species. Nor even in the warmer parts of Europe is it partial to the moist atmosphere of the Atlantic seaboard.

As long as the Military Orchid is still with us, the last week of May or the first week in June seems the best time to admire it.

IIj. *Orchis simia* The Monkey Orchid

* 11th June Kent

If the Military Orchid resembles a rabble of toy soldiers, this plant looks like a cascade of monkeys on strings with legs kicking and arms flailing, with much thinner arms and legs than those of the Military Orchid. The Monkey Orchid has no 'metallic' helmet, but instead a white or mauve one with the sepals faintly dotted with violet and not so tightly packed into the helmet.

This orchid is peculiar in that the flowers at the top of the spike open before those lower down, which helps to account for the fact that the spike, instead of resembling a spire, is 'flat-roofed'. Nevertheless, once the orchid is in full bloom, the upper flowers are so closely packed that they seem still to be in bud while the lowest flowers, having more space to themselves, look more fully developed. The photograph chosen shows the mauve variety of the orchid known as var *macra* Lindl.

A century and a half ago, this plant was fairly common on the chalk hills of the Thames Valley between Pangbourne and Henley. 'On Cawsham (Caversham) Hills by the Thames side, not far from Reading in Berkshire' is the locality quoted by John Ray in 1724 for this plant and until about 1840 it was still plentiful near Pangbourne. Since then it has decreased in numbers partly because of turf removal, collecting and rabbit grazing, and it has become a very rare plant.

Only two sites in Oxfordshire survived into the 1980s although at one of these more than thirty spikes were recorded in 1933. Unfortunately the site, though studded with bushes, was near a footpath and the orchids were regularly picked. Furthermore the slope on which they grew was not steep enough to protect the site from the plough. Some plants however were rescued and moved to a steep escarpment nearby where they (or seed from the plants at the original site) have survived. A second Oxfordshire site was found in 1966.

Kent is the other county in which the Monkey Orchid survives, amid some controversy. At one time in the 1920s a small colony existed near Bishopsbourne to the south-east of Canterbury. They disappeared. Then thirty years later a flower appeared on the vicarage lawn at Otford, near Sevenoaks. The plant bloomed regularly and, with praiseworthy optimism the seed was broadcast on likely patches in the neighbourhood. Following the appointment of a new vicar less concerned with flowers of the field, the orchid was removed 'for safety' and replanted elsewhere in rough ground. As might have been expected the plant prospered for a year or two and then died. Nor have any of its offspring appeared since on the vicarage lawn. In addition, however, a small colony with a varying number of plants was discovered on a chalky bank close to Ospringe near Faversham and other plants have turned up occasionally in the same neighbourhood. During good years seed has been taken from the Ospringe colony and sown in a number of different localities including the Kent Trust for Nature Conservation at Park Gate, where a new colony seemed at one time to have been successfully established. Clearly this development took the pressure off the original plants at Ospringe but met with disapproval of the compilers of the *Atlas of the Kent Flora* which commented: 'This sowing of native seed of native species whilst very commendable in botanic gardens is to be deplored over the countryside in general. Not only does this weaken the conservation lobby (presumably because it suggests that the native colony is not irreplaceable), but it will completely invalidate any further finds of this species as being native plants.'

Meanwhile for the next chapter in the saga of the Monkey Orchid we have to turn northwards. In 1974 a Mrs A. Fritchley, while on a visit from Leeds, discovered a single plant of the Monkey Orchid on the narrow spit of the Spurn Peninsula Nature Reserve in S.E. Yorkshire. It was flourishing there amid the dunes – a habitat which, because of the calcium content of the sand, suits many orchids. Quite possibly the seed may have arrived there in the chalk imported for strengthening the ridge of the spit. At first only a single plant was found; then more appeared, and by cross-pollination the colony was increased to as many as thirty plants. Then, alas, in 1983 disaster struck. During a heavy storm the sea broke over the site covering it with salt, and destroying the colony. There is a local belief that because of coastal erosion the spit of Spurn

Head is moved some distance inshore in a cycle of about 250 years. One can but hope that when the coast is stabilised for the next 250 years, the new Spurn Head Reserve will be built up again with equally productive chalk.

IIk. *Potentilla fruticosa* Shrubby Cinquefoil

Surely the most delightful of all settings for this plant must be the stretch of the river Tees below the great waterfall of High Force, near to Bowlees, where the Durham County Conservation Trust has an information centre. From there, with permission, you may walk across a couple of fields to a small wood and, through it, to a miraculous suspension-bridge built for pedestrians only (and one at a time) over a deep gorge of the river. Winch Bridge, as it is called, leads to an island, and a great deal more – a panorama of waterfalls and rocks over which the wary can step without too much danger.

On the bank of the island, upstream from the bridge, and in most cases safely out of reach of collectors (it is, in any case, a very public beauty spot) is *Potentilla fruticosa*, the Shrubby Cinquefoil. *Potentilla* is a name bequeathed to us by the apothecaries, who believed that other members of the same family, if not this one, possessed great potency. The *fruticosa* adjective is not, as might have been supposed, connected with fruit or fruitfulness, but is the Latin for shrubby – which is what this plant is.

What a pity it is that it does not flower at the same time as the Spring Gentian which grows not so very far away! Instead, the end of June is a suitable date to mark in the diary. Nearby, and a minute or two's walk from the bridge there are other splendid plants – *Antennaria dioica*, a neat little woolly-white mountain species, almost unknown to visitors from the south, and *Prunus padus*, Bird Cherry, with its spires of white flowers, again uncommon – unless planted – south of the Thames.

Shrubby Cinquefoil grows to as much as a metre in height but often keeps a lower profile. The leaves, like those of its relatives, the roses, are pinnate, that is they are composed of a number of leaflets arranged in two rows in opposite pairs along the stalk. Apart from the leaflet at the end of the stalk, there are usually two pairs of leaflets to each leaf, making five leaflets in all. But sometimes there are three pairs, making

IIa.

IIb.

△IIc. ▽IIe.

IIf.

IIa. *Draba aizoides* (ht 10 cm)
IIb. *Dianthus gratianopolitanus* (ht 20 cm)
IIc. *Dryas octopetala* (fl. diam. 3 cm)
IIe. *Gentiana verna* (fl. diam. 2 cm)
IIf. *Gladiolus illyricus* (ht 90 cm)

IIg.

IIh.

△**IIi.** ▽**IIl.**

IIj.

IIg. *Helleborus viridis* (ht 30 cm)
IIh. *Melittis melissophyllum* (ht 40 cm)
IIi. *Orchis militaris* (ht 30 cm)
IIj. *Orchis simia* (ht 20 cm)
IIl. *Pulsatilla vulgaris* (ht 30 cm)

IIn.

IIp.

△**IIq.** ▽**IIt.**

IIs.

IIn. *Saxifraga oppositifolia* (ht 15 cm)
IIp. *Veronica fruticans* (ht 15 cm)
IIq. *Tulipa sylvestris* (ht 60 cm)
IIs. *Myosotis alpestris* (ht 10 cm)
IIt. *Linnaea borealis* (ht 10 cm)

IIIa.

IIIb.

△IIIc. ▽IIIf.

IIIh.

IIIa. *Adonis annua* (ht 35 cm)
IIIb. *Agrostemma githago* (ht 90 cm)
IIIc. *Bupleurum rotundifolium* (ht 30 cm)
IIIf. *Silene noctiflora* (ht 60 cm)
IIIh. *Cirsium tuberosum* (ht 90 cm)

up a total of seven leaflets. The plant is easily distinguishable from other *Potentillas* in that all its leaflets are 'entire', that is not slashed or cut.

Apart from Upper Teesdale, the plant has been recorded in the Lake District (Wastwater), as well as on Helvellyn, from Yorkshire, Westmorland and Cumbria almost always on damp 'basic' rocks and in Ireland in the rocky area known as the Burren, where, however, it shows a decrease.

The flowers are yellow, sometimes nearly an inch in diameter, but few in number at any one time, which luckily makes it less attractive to predatory gardeners. In any case, many 'improved' varieties are obtainable from nurserymen.

III. *Pulsatilla vulgaris* Pasque Flower

* 8th April Gloucestershire

How much at risk is the Pasque Flower? It is not yet mentioned in the *British Red Data Book* and, at present, is to be found in a band of country stretching almost from the Wash to the Severn on chalk and limestone. But it remains an extremely local plant. The Sussex Downs are not to its taste, nor those of Surrey or Kent which shelter so many other rare chalk- loving plants, although the Berkshire Downs satisfy its requirements.

Through the kindness of a farmer I was able to see Pasque Flowers in Gloucestershire in what is probably its most westerly station. Crossing a Cotswold stone wall, one is on the billow of a down and, after crossing another wall, one is on another billow among the clouds, shielded from all motorists and where the viewer is free to enjoy the plants in privacy with the sheep at a time when the grass is still short. At first all that can be seen – for it is still early in April – are threads of the silken feathery calyx-like structure (the involucre) and the stem leaves that surround the bud. The six, deep purple sepals which compose the flower are also coated underneath with silk and have not yet started to glow. When they first show, the buds are still ground-hugging like violet pebbles. The flowers are more easily distinguished as the plants begin to thrust up, and the blooms are upheld well above the stem leaves on stalks which grow while they blossom. As the sepals start to open, like the leaves of a compur shutter in an old-style camera, they disclose a quiver of golden stamens, those in the centre having fertile anthers while those round the edge carry nectar but are infertile.

At this point the flower is bell-shaped but soon the sepals will expand to present a six-pointed purple star before beginning to turn back under. All this should have taken place by Easter Sunday for Pasque is a variation of the Latin-based word for Easter. Even in seed it remains beautiful for the flowers droop becomingly, ready to shed the achenes, which, like those of their relatives the clematis family, are plumed.

The Gloucestershire site is on limestone of the kind known as oolite from the Greek for 'egg' – because the soil is composed of granules of the kind to be seen in the hard roe of fishes. As I turned away from the oolite down for the road, I wondered what would happen to this farm in the future. Would the same farmer and his family continue to manage it? Would the sheep be turned off it to make way for the tractor? Would the rabbits take to eating Pasque Flowers? Would there be houses such as we see on the Downs above Brighton and would our children and grandchildren be surprised to learn how widely distributed the Pasque Flower once was as we are today when we read, for example, about the Monkey Orchid? Still while we live, the Pasque Flower may yet be found nearer to London than Gloucestershire – in Cambridgeshire, Bedfordshire and, the best known site, near Royston in Hertfordshire. A bulb of *Pulsatilla vulgaris* can, however, be bought from nurseries for the current price of one pound: less than the cost of motoring fifty miles.

IIm. *Saxifraga cernua* The Drooping Saxifrage

Here is a botanical curiosity which some gardeners, unfortunately, have taken to their hearts. It stands less than 15 cm (6 ins) high, with bright green leaves, kidney-shaped at the base of the plant, and ivy-leaved on the stem. One single flower – sometimes no more than a bud – tops the end of the stem which curves over like a shepherd's crook, as if the flower were too heavy by half.

This Saxifrage seldom grows below the 3,000 ft level and is quite happy at 4,000 ft. Understandably, with the flowers seldom open and not many insects to be found at these levels, the chance of normal pollination seems remote. But the plant reproduces itself from bright red bulbils – small bulbs – which form in the axillae of the leaves.

That invaluable guide, *Ben Lawers and its Alpine Flowers*, published by the National Trust for Scotland, tells us that the lowest level at which the plant is to be found there is 3,600 ft, but adds it is also found on the

Ben Nevis range and on Bidean nam Bian, Glencoe. Considering the way in which this Saxifrage likes to hide behind large Stonehenge-type boulders in the remoter parts of the Highlands, there may be other as yet undiscovered sites in Perth, Inverness or even Argyll. If you find a plant of the Drooping Saxifrage, as I've said, you are likely to see, near to it, a coloured pinhead, like those that some military men use on their maps. These pins help researchers to record the numbers of plants each year and their survival rate. As I looked at the array, I could not help recalling the circumstances in which the Reverend W. Keble Martin, author of the best-selling *The Concise British Flora In Colour* visited Ben Lawers in order to make drawings of the Saxifrages there.

After a full Sunday's work at his parish of Coffinswell, near Newton Abbot, he caught the midnight train to Killin and on to Ben Lawers (no station there now). The next morning found that the clouds were down to about 1,000 ft, the point at which the climb really starts. Nevertheless, he walked up into the darkness and rain and found two 'nice' Saxifrages, including one particular one which, he said, was very local and flowered 'sparingly'. This must surely have been *Saxifraga cernua*, the Drooping Saxifrage. The date, 25th July, should have been just right. He took the plant down, made a drawing and coloured it. (He could have made the bulbils larger and a more brilliant red than in the book). Next morning he repaired to the top of Ben Lawers once more, climbing again through the mist and rain, found the niche from which he had taken the Saxifrage and replanted it firmly before descending with some more Alpines, presumably less valuable, which he carried away in cigarette tins for drawing in the train on the way home.

Keble Martin's practice was to make preliminary drawings on scraps of thin paper. These were then traced on to drawing paper laid on a sheet of glass. Eventually 1,486 species of plants were illustrated in colour in this way. But, in some ways the more limited monochrome drawings published in Keble Martin's *Sketches for the Flora* are just as vivid.

These were the roughs afterwards adapted by the artist to fit the page they were to decorate. Underneath them are the botanical and popular names, hand-written with the dates and locations, and an indication of whether the plant was drawn in the wild, or based on a dried specimen in a herbarium. Those who would like to follow the same practice with Keble Martin's sketch book will be pleased to know that the paper was specially chosen to allow flower portraitists full freedom of expression. In the last century and the early years of this, it was the custom of botanists to hand-colour W.H. Fitch's line illustrations in Bentham and Hooker's *Handbook of the British Flora*.

It is intriguing and encouraging to be able to confirm from one's own experience that some of the rare plants that Keble Martin drew continue to appear on the same date – more or less – and at the same place.

His *Cypripedium calceolus*, Lady's Slipper Orchid, was however drawn from a specimen in the British Museum – after he had 'waited many years for news of a flowering specimen'. The final colouring was based on fresh flowers from a garden in Lancashire though the petals are of a deeper colour and purpler than those of the wild plant.

It is an awe-inspiring thought that the first drawing for Keble Martin's book was executed in the nineteenth century when he was about twenty-two. It took him some fifty years to complete the work and it was published only four years before his death in 1969 at the age of 92.

IIn. *Saxifraga oppositifolia* Purple Saxifrage

* 18th April Powys

Seldom rising more than one inch above the ground, the Purple Saxifrage is a humble member of a large family which includes such magnificent garden plants as *Philadelphus*, the Hydrangeas, the Escallonias and, for the housewife, currants and gooseberries. But the Purple Saxifrage flowers at the end of the branches, with the petals just over 1 cm (almost ½ in) long, are large for the size of the plant. Rosy-purple rather than true purple is probably the right description for their colour.

The leaves are dark-bluish green fringed with hairs oval or obovate (broader nearer the tip than in the middle) and closely crowded together, in four rows. At first the plant looks as though it had been showered with white dust, but a close examination of the leaves reveals that each is flattened and truncated near the tip, and that, in the flattened area, is a pore from which small nuggets of lime are expelled. One would hardly have expected this to happen, as the Purple Saxifrage favours sites that are rich in lime. But the plant also likes constant running water and perhaps this sometimes contains more lime than was bargained for. At any rate most Purple Saxifrages seem to have lime to spare. This curious pit, or pore, is a feature of another much rarer plant *Scheuchzeria palustris*, the Rannoch Rush which is found mainly on Rannoch Moor near Glencoe. This, like the Purple Saxifrage, is very much a north country plant but it is to be found in sphagnum bogs which are acid and the very opposite of the kind of place chosen by the Saxifrage. The purpose of the Rannoch Rush's pore is not too clear, but it has not brought about any real increase in population.

Traces of the Purple Saxifrage have been found in Britain in deposits that were laid down 20,000 years ago, that is before the end of the last Ice Ages. As each of the glaciers retreated north towards the Pole, the Arctic alpine plants – of which the Purple Saxifrage is one – followed up, going ever northward. The Arctic alpines were in turn followed by plants that could live at higher temperatures and, when the ice finally vanished for good, the Saxifrages and other Arctic alpines had to find refuge wherever they could on high mountains, cliffs and the like.

For the Purple Saxifrage there is as yet no shortage of sites. It is to be met with in the Lake District on ledges and screes, in the Yorkshire Dales on crags, in Snowdonia in the Hanging Gardens, on Ben Lomond, Ben Lawers, in Glen Caenlochan and, for southerners, on cliffs in the neighbourhood of Brecon, where, in a normal season, it will still be in flower in April.

The risks to this attractive plant are threefold. First, some alpine gardeners fancy it (although improved varieties can be found in good nurseries). Secondly, it is eaten by sheep and on Snowdon by wild goats in any place that can be reached, and thirdly, any interference with its water supplies – as for instance through pollution or trampling – can prove disastrous. Occasionally it is washed away downhill by floods.

IIo. *Stachys germanica* Downy Woundwort

At first sight this stately member of the Dead Nettle family looks like a garden escapee of *Stachys lanata*, popularly known as Lamb's Tongue. Both plants are covered with white silky hair and are decked with whorls of purplish pink lipped tubular flowers. But *S. germanica* is unbranched, and grows to 80 cm, (say 2 ft 6 in) whereas *S. lanata* is mat-forming and rarely more than 40 cm in height. The leaves also smell somewhat foetid. *S. germanica* which formerly grew in a half-dozen counties is now confined to an oolite limestone area in Oxfordshire lying between Oxford city, Burford and Chipping Norton. Stony grassland and even roadside verges are favoured, and the plant, both conspicuous, attractive and accessible is considered at high risk, even though the garden species offers a nearly equivalent substitute. *S. germanica* is sometimes regarded as a short-lived perennial, sometimes as a biennial flowering in its second year of growth and then dying away. The number of plants – and consequently of flowering plants – also varies from year to year according to local conditions, often increasing after competing growth has been cut away.

IIp. *Veronica fruticans* Rock Speedwell

* 6th July Tayside

The Rock Speedwell is one of the most attractive of Britain's smaller flowers. It is not enough to say that its petals are blue. Other Speedwells have blue petals. Those of the Germander Speedwell, *Veronica chamaedrys*, are deep sky blue; those of *Veronica polita*, the Grey Field Speedwell, are true-blue as are those of *Veronica arvensis*, the Wall Speedwell. But those of the Rock Speedwell have in them a tinge of crimson, making the flower somehow less impersonal, more communicative: the difference between a real eye and a glass one. Each bloom has within it a scallop of crimson, circling the whole of the inside of the corolla, and within the crimson circle between that and the centre of the flower are four pips, one at the base of each petal, arranged in cruciform pattern, like a quadrifoil in a church window. At the very centre of the bloom, like a jewel, stands a golden carpel. Seen at their best, the flowers are the enamel in a fresco of fresh green leaves that stream like a torrent down the side of some rock. All too rarely alas. Rock Speedwell are to be found in no more than a score of localities in Scotland, in counties that were once called Perthshire, Aberdeenshire, Angus, Inverness-shire and Argyll. It is also one of the features on Ben Lawers on Tayside, known as the Botanical Mecca for alpine flowers. So let us stay for a moment or two, and consider why the rocks there of Ben Lawers are so special.

The Rev. John Lightfoot (1735–1788) whose *Flora Scotica* was published in 1788 learnt of the alpine flowers in the mountains around Ben Lawers through the Rev. Mr Stuart, Minister at Killin, but James Dickson (1738–1822), a Covent Garden nurseryman, seems to have been the first collector to have explored Ben Lawers itself in detail. His herbarium of dried plants went to the Linnean Society of London and is now at the British Museum (Natural History) at South Kensington. From this it is clear that he paid two visits to Ben Lawers, the first in 1789 and a second in 1792 during which he discovered both *Saxifraga cernua*, the Drooping Saxifrage, the peculiarities of which have already been discussed and *Gentiana nivalis*, the Alpine Gentian.

Ben Lawers today is under the protection of the Scottish Naturalist Trust which was enabled to purchase the south side of the mountains of Ben Lawers, and Ben Ghlas, to the south-west of it, in 1950 through the generosity of the late Mr P.J.H. Unna. Mr Unna had been an enthusiastic supporter of the Scottish Mountaineering Club and it was his ambition that as much as possible of the Scottish mountainside should be preserved unspoiled and freely accessible to all. Mr Unna, who died in a climbing accident on the face of Ben Aighennan near Dalmally in 1950 a few months after the acquisition of Ben Lawers, left the whole residue of his estate to the Trust.

If he were alive today he would, I am sure, be pleased to see how well the Trust has interpreted his wishes. An information centre, built of local stone and dark timber, has been set up near the foot of Ben Lawers to help visitors in every possible way, and for those who can stay only a short time there is a Visitors' Nature Trail which takes them up a highland burn luxuriant with Alpine Lady's Mantle and Mountain Yellow Saxifrage to name just two outstandingly beautiful plants.

Geologists tell us that Ben Lawers is built not of rocks at one time molten in the earth's heat, but of fragments of soil and rock deposited on the site as sediment. Later, in the era when great mountains were being formed, this sedimentary material was compressed concertina fashion into folds, and, in some areas, the layers containing minerals were compressed to form brittle plates known as schists which shatter easily along parallel lines. The schists at Ben Lawers yield supplies of minerals essential for a rich flora, namely salts of calcium, magnesium, sodium, potassium, iron, as well as phosphate and sulphate. They are near the summit, exposed to wind, and to frost, which fractures the schists – and to rain which washes the mineral particles downhill across the rock-faces and on to the ledges. The rain would seem to be plentiful enough to ensure a constant supply of minerals to the plants lower down the slope but not so heavy that the minerals are washed away completely. And so each year, as the rock above erodes, the alpine and arctic plants of Ben Lawers are granted a new lease of life. In places where water seeping down internally encounters impervious rock it wells to the surface in what are known as wet flushes. It is in such spots as well as on ledges where the sheep cannot climb that the finest rarities, including the Rock Speedwell, are to be seen.

IIq. *Tulipa sylvestris* Wild Tulip

* 29th April Middlesex

One might have supposed that a wild tulip would be a small under-privileged version of the garden plant, like a wild strawberry in comparison with those served at Wimbledon. But no. There was no need to improve this plant which, when growing in a copse or shady wood, will rise to 61 cm (2 ft) though 46 cm (18 ins) may be confidently looked for. Tulip enthusiasts grow it in their beds, despite the fact that it sometimes falters after a couple of seasons. It is easily distinguished from other varieties or species of tulip by its three reflexed outer petals, each striped beneath with green. The buds, pointed like the nose of a projectile, hang down at first and the three inner petals await the sunshine of April before disclosing their glory. The leaves also, usually three in number, are slightly glaucous.

The Wild Tulip is presumed to be an introduced plant because the orchards and woods in which it appears are often relics of earlier gardens or are connected in some way with a local stately home. Yet, though not prolific, *Tulipa sylvestris* has been recorded in about one-third of the vice-counties. It has established itself in its own right as a naturalised species both in the east of England and in Somerset, Gloucestershire and Wiltshire. In Scotland it is found as far north as Fife. In Essex it turned up unexpectedly on the Leigh Marshes, a salt-marsh nature reserve watched over jointly by the Essex Naturalist Trust and the Nature Conservancy Council. Land there had been reclaimed mainly in order to encourage sea-birds to take up station there, partly by bringing in new soil and tipping it out. Somewhere, somehow, a wild tulip was included in one of the skips; and it flowered. A better known and more dependable site is at Harefield, Middlesex, not far from the famous heart hospital and London's Heathrow. It is wardened by the local Natural History Society.

IIr. *Phyteuma tenerum* Round-headed Rampion

From a distance this flower gives a general impression of an outsize deep violet clover. It is in fact a member of the Bellflower family, though the flowers that make up the clover head are more tubular than bell-like in the generally understood sense. At a later stage the five petals which form the tube separate from one another and spread out down to their common base. The lower leaves are scalloped, heart-shaped and stalked; the upper leaves unstalked.

This is one of the protandrous plants; that is the stamens mature before the ovary, and the pollen is presented to visiting insects on the twin stigmas of the style as it grows upwards.

Phyteuma from the Greek *Phyton*, a plant, was the name given supposedly to this one by Dioscorides in the 1st century A.D. Rampion, first used in English in 1573, was derived from mediaeval latin words used for several vegetables including *Campanula rapunculus*, Rampion Bellflower, the leaves and roots of which were eaten in salad. (It corresponds to the French *raiponce* which the *Larousse Gastronomique* tells us yields a root which can be eaten either raw or cooked. In the latter case the roots are scraped, then divided into pieces of convenient size, dipped in water to which a little vinegar or lemon has been added, and cooked gently in a flour and water mixture for about two hours. The leaves can be eaten raw in salad or cooked like spinach.)

Whereas *C. rapunculus* is happy on a gravelly soil, *Phyteuma tenerum* is a plant of the chalk grasslands and is found in quantity only in the southern counties ranging from Dorset and Wilts through Sussex to Kent. Coming on a clump of them for the first time is quite an experience.

IIs. *Myosotis alpestris* Alpine Forget-me-not

✻ 8th July Ben Lawers

The most remarkable feature of this plant – which is sometimes little more than 5 cm (2 ins) high – is the comparatively large size of its bright blue flowers – about one-third of an inch across. This is, in proportion, about twice the size of the flowers of other Forget-me-not species such as *Myosotis arvensis*, the Common Forget-me-not, and *M. sylvatica*, the Wood Forget-me-not. So it is all the more exciting to find this plant, looking like a cultivated hothouse version of the 'ordinary' Forget-me-nots in its typical habitat – a cranny on the bare mountainside, as for instance on Ben Lawers.

But the photographer, if he wishes to carry away a remembrance of such beauty, must not focus his camera on the flowers from above. For in this case, nearly always, the film will record the greyish or whitish light of the sky, reflected upwards on to the film by the corolla of the flower, and almost all the blue will have vanished. Instead a deep obeisance is required so that the petals can be glimpsed from below, with the light shining through them.

Apart from its beauty and rarity, this Forget-me-not, as its name implies, is distinguished as one of the few of the UK mountain plants to be found also in the Alps, but not in the arctic regions. It shares this distinction with *Gentiana verna*, Spring Gentian, and *Viola lutea*, the Mountain Pansy, although these last two are denizens of the uplands rather than of mountains. Rather more of our mountain flowers, including the spectacular *Chamaepericlymenum suecicum*, Dwarf Cornel, and *Erigon borealis*, Boreal Fleabane (sometimes inaccurately Alpine Fleabane) are arctic-based and are not to be found in the Alps. Many others, however, including *Dryas octopetala*, Mountain Avens, are at home wherever the climate is right whether in the Arctic, the Alps or even the Pyrenees.

IIt. *Linnaea borealis* Twinflower

* 9th July Braemar

Of all the many plants that he named, the great Linnaeus acclaimed this as his own. He fell in love with it soon after his twenty-fifth birthday on a Sunday afternoon in May 1732 while on a journey to Swedish Lapland, a trip sponsored by the Royal Swedish Society of Science. He travelled alone on horseback, in a cloth coat and second-hand leather breeches which he had bought at auction, carrying a bag containing amongst other things a shirt, two pairs of half-sleeves, two nightcaps, an ink-horn, a pen case, a stock of folio-size paper for pressing plants and a gauze veil to keep out the midges. Linnaeus never hunted for plants on a Sunday until after he had attended church service, and so it was not until the afternoon, when he had passed the city of Gaville, he came upon a plant which was then called *Campanula serpyllifolia* – the Thyme-leaved Bellflower – a name given to it by Caspar Bauhin of Basle in 1596.

Linnaeus was right to reject this name for this lovely plant belongs to the Honeysuckle family, and it was at the suggestion of his friend Gronovius, no doubt with Linnaeus' joyful concurrence, that it became Linnaeus' name-plant. Linnaeus in his work *Critica Botanica* wrote: 'This plant was named by the celebrated Gronovius, and is a plant of Lapland, lowly, insignificant, disregarded, flowering for only a short time – after Linnaeus who resembles it.'

Linnaeus was perhaps exaggerating somewhat if he was suggesting that the flowering season of Linnaea is more quickly over than that of other flowers. No doubt this depends on the latitude and other local conditions. At Braemar, which weathermen often cite as the coldest place in the U.K., it is in bloom in mid-July. But June and August have also been recorded.

The bell-shaped flowers, hanging each on the end of the finest of long upright stalks, vary in colour from rose-pink to white, tinged pink, and, as our photograph shows, the inner side of the petals looks as though marked with a dab of raspberry jam. The leaves are small, toothed and oval, but once the leaves are past, the plant creeping beneath the blaeberries, whortleberries and Ling becomes almost invisible.

Linnaea is one of the refugee plants that had to move northwards when the glaciers of the Ice Age melted and the broad-leaved oaks and elms began marching northwards to replace the ancient pine forests. It was once to be found in Yorkshire and in the lowland woods of Scotland. When Professor Beattie of Aberdeen saw the plant in 1795, it grew in the pinewoods of Mearns, a district now in the shadow of Glasgow. It is now very local and rare and is now to be found mainly in the Highlands in the east of Scotland in places sheltered from the sun either by rocks or tall trees of the remaining pine woods.

IIu. *Dianthus deltoides* Maiden Pink

John Gerard wrote about this pink: 'There is a wilde creeping pink which groweth in our pastures neere about London, and in other places but especially in the great field next to Detford, by the pathside as you goe from Redriffe to Greenwich, which hath many small tender leaves, shorter than any of the other wilde Pinkes, set upon little tender stalkes which lie flat on the ground, taking hold of the same in sundry places, whereby it greatly increaseth; whereupon grow little reddish floures. . . .' To Gerard, then, this red-floured Pink was the Deptford Pink.

He then went on to describe another Pink with flowers 'of a blush colour, whereof it took its name, which sheweth the difference from the other'. Gerard knew this as the Maidenly Pink. But somehow the captions to the illustrations in Gerard's *Herbal* became transposed. The drawing, one of the sharp red pink was titled Maidenly Pink and the other plant with the blush pink flowers became the Deptford Pink. And so it has been ever since.

The flowers of *D. deltoides* are distinguished by a fretted band of darker red around the centre, which sometimes takes the form of a pentagon or of broken lines of dark red dots. Irregularly placed white spots are sometimes present outside the ring. The leaves are often glaucous green.

This is an uncommon and local plant though sometimes plentiful where found. It favours dry banks and pastures (including golf courses). The plants are sun-loving and the petals close towards sunset or when black clouds threaten, making them hard to detect. *D. deltoides* can be bought from nurserymen and will sow itself into crevices of the rock garden or of the right type of wall.

— III —

Aliens among the Corn and Elsewhere

* (See following p. 82 and facing pp. 98 and 99 for illustrations)

IIIa. *Adonis annua* Pheasant's Eye

* 6th August East Sussex

The wild plants that grow on arable land are more at risk in Britain than any others to judge by the numbers of those extinct or likely to become so. Interesting and attractive plants which would normally survive and even flourish in disturbed ground fail to do so when subjected to weed-killer and stubble-burning.

Pheasant's Eye is one of the most beautiful of the plants in this group. It is the equivalent of a miniature paeony, with blood-red petals, each with a dark spot at the base. They say that the name was bestowed in memory of the youth Adonis whose blood stained the petals as he met his death from a wild boar.

Our Pheasant's Eye used to be called *Adonis autumnalis* to distinguish it from *Adonis aestivalis* which flowers earlier in the season, but the latter appears so seldom, and the 'Autumn' Pheasant's Eye flowers in July and August in Britain (if not in Linnaeus' Sweden), so that the adjective 'annual' is now used in the botanical name of our plant. Our Pheasant's Eye bloom is more or less bowl-shaped, often with one of the petals a miniature version of the others, a peculiarity to be seen in one other member of the same family – *Ranunculus auricomus*, the Goldilock's Buttercup.

The sepals beneath the Pheasant's Eye are sometimes green, sometimes purple and they stand away from the petals – a refinement which distinguishes *R. annua* from *R. aestivalis*. The leaves of both are finely divided, feathery green – almost hair-like, and the fruits are arranged in formation which suggests a miniature green loganberry or even a pineapple.

Adonis annua is said to be an introduced plant and therefore does not merit inclusion in the *Red Data Book*. (It would, in any case be difficult to protect.) But it has nevertheless become naturalised in some southern counties from Dorset to Oxford and Kent – from which one may assume that one should look for it in chalk areas. Sir Edward Salisbury in his work *Downs and Dunes* (1952) wrote that Pheasant's Eye 'was once so common in the cornfields of Sussex . . . as to have been collected and sold in Covent Garden Market'. As late as 1978 a ploughed field near the sea at Crowlink in East Sussex which had been allowed to run wild produced a display which reddened the field for a hundred yards or more.

IIIb. *Agrostemma githago* Corncockle

✻ 23rd July Hampshire

In the early 1920s one could be almost certain of seeing one or two of these splendid purplish pink flowers, up to 5 cms (2 ins) across, in the average cornfield. Plants almost a metre tall were not uncommon, each with a branch or two and a single bloom at the end surrounded by a collar of spiky green sepals. 'Much too common' was the general verdict among farmers as the seeds of Corncockle, large, black and unpalatable, were a problem for flour millers and discounted the value of the crop.

However, in the years between the two World Wars, improved sorting methods made it possible to separate out the seeds of Corncockle from the seed-grain imported into Britain from Europe, and, from then onwards, our cornfields became more productive but less interesting.

During the period 1930–1960 (which provided the basis for the *Atlas of the British Flora*), Corncockle was recorded in fewer than 200 of the 10-kilometre squares into which the country was divided, and since the Fifties its appearances have been rare. Our picture here was taken on one of the development plots of the Butser Hill Farm in the Queen Elizabeth Country Park near Petersfield, and a number of introduced plants have survived for some years in the corner of a cornfield near Swaffham Prior not far from Newmarket. Apart from these two sites, the best chance of seeing Corncockle is around the nearest chicken-farm where imported grain may not have been so scrupulously sorted.

IIIc. *Bupleurum rotundifolium* Thorow-wax

✻ 23rd July Hampshire

The English name, Thorow-wax, is well chosen, for its stalks do seem to grow through the bluish leaves. At least through the upper leaves, for the lower ones narrow into the stalk. This is a member of the Umbelliferae – the umbrella-bearing family in which all the flower stalks rise from the same point on the main stem. Most members of this family – such as Wild Carrot, Parsley, Hemlock, Fennel, Angelica etc. – have deeply divided leaves and white flowers, and are often distinguishable from one another with certainty by minute differences between their exceedingly uninteresting 'fruits'. The genus *Bupleurum* however not only has simple, undivided leaves (which have led to their being given the alternative name of Hare's Ear) but are distinguished by yellow flowers.

But at this point another difficulty arises. *Bupleurum rotundifolium* which was once a common cornfield weed in the south-east of Britain is now thought to be extinct. But another Hare's Ear, very like it in appearance, keeps appearing, and making the heart of the amateur start to pound, thinking that he or she has re-discovered an extinct species. The look-alike plant is *Bupleurum subovatum*, also an introduced plant that probably was imported with bird seed – though the specimen I found was on the edge of a cornfield in Sussex where I was on my way to look for the Nottingham Catchfly – and miles from any budgerigar.

The leaves of *Bupleurum subovatum* (sometimes *lancifolium*) are not, in fact, lanceolate but rather ovate, that is with the widest part of the leaf below the middle. The leaf of the 'extinct' plant is elliptical, that is the widest part is in the middle. The extinct plant may have as many as ten rays in its umbrella, the bird-seed plant often has but two or three.

The photograph in this book was taken at Butser Hill Farm near Petersfield where rare cornfield plants are grown from seed.

IIId. *Ajuga chamaepitys* Ground-pine

As little as two inches in height, this plant which belongs to *Labiatae* or Dead-Nettle family must often escape notice. It not only looks like a small pine seedling, but is said, also, to smell of pine leaves. The flowers, arranged in pairs, are partly concealed amid a pyramid of dark green leaves. They are yellow with red spots on the lower lip.

According to *The Flora of the British isles* by A.R. Clapham, T.G. Tutin and E.F. Warburg (1962 edition), it occurs in only nine vice-counties. Its most favoured station is on banks, chalkpits, or by the edge of arable fields or the tracks around them, but, according to J.E. Lousley, it will also appear in clearings in woods. The same writer points out that it has appeared where rabbits have been active, where chalk has been dug up for pipe-laying or in gardens that have been dug. The north Kent Downs are a favourite haunt, although, even in John Ray's time, centuries ago, it was not common. Since then no doubt path-widening, spraying, 'development' and the attacks of myxomatosis which has reduced the rabbit population, have taken their toll.

IIIe. *Verbascum lychnitis* White Mullein

A chalky bank, an old (chalk) quarry or a railway embankment would be the most likely spot for this Mullein, and north-west Kent offers the best hunting ground, although it has been reported within an area ranging from Devon in the west to north Wales and East Anglia. Most Floras describe it as 'very local' and, at one time, it appeared to have vanished from a well-known site on the Sussex Downs near Arundel only to reappear in 1986 in the same area after the ground there had been disturbed.

There are two varieties of this species, one with white flowers and the other with yellow, but as the English name suggests the white variety is the more common, and the yellow sort is found only in Somerset. In both, however, the leaves are dark green above but densely covered with starry tufts of short soft white powdery hairs. The filaments (the stalks of the anthers) are covered with white hairs. The stem is angled and densely powdered like the undersides of the leaves.

The Figwort family (Scrophulariaceae) to which the mulleins belong contains many plants such as Foxgloves, Snapdragons, Cow-wheat and others described by botanists as zygomorphic in as much as the flowers appear symmetrical (capable of being divided into two or more exactly similar and equal parts) only if seen from one particular viewpoint. Foxgloves, Snapdragons and Cow-wheats are strongly zygomorphic; mulleins and Speedwells (which also belong to the same family) are less markedly so, and it has been suggested that these latter are less 'advanced' than the complete zygomorphs.

On the European continent the White Mullein grows as far north as Belgium and northern Germany but it is more at home in central Spain, Italy, Greece, and Morocco and it may be supposed that our fickle climate has something to do with the fact that it is not more widely seen.

IIIi.

IIIj.

△IIIk. ▽IIIo.

IIIn.

IIIi. *Teucrium chamaedrys* (ht 15 cm)
IIIj. *Orchis ustulata* (ht 20 cm)
IIIk. *Scandix pecten-veneris* (ht 25 cm)
IIIn. *Melampyrum arvense* (ht 40 cm)
IIIo. *Melampyrum cristatum* (ht 30 cm)

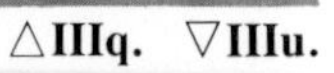

△**IIIq.** ▽**IIIu.**

IIIr.

IIIt.

IIIq. *Salvia pratensis* (ht 100 cm)
IIIr. *Orchis purpurea* (ht 40 cm)
IIIt. *Gagea lutea* (ht 15 cm)
IIIu. *Silene otites* (ht 50 cm)

IIIf. *Silene noctiflora* Night-flowering Catchfly

* 4th July Hampshire

At times one might think that it is a rather weak version of *Silene alba*, the White Campion, which is common throughout Britain on fields, cultivated land and by hedgerows. Both are scented at night and both have whitish flowers; the Catchfly with little more than a hint of pink, the Campion a good deal pinker if, as frequently happens, it hybridises with its cousin the Red Campion. The main difference between them is that in the light of the day and the heat of the sun, the petals roll inwards – something that never happens in the case of the commoner plant. The undersides of the Catchfly petals when rolled up are seen to have a yellowish cast.

One other plant for which the Night-flowering Catchfly could be mistaken is *Silene nutans*, the Nottingham Catchfly, the petals of which are also rolled up during the day, but as 'nutans', the botanical name for the species, correctly implies, the flowers do nod downwards and are generally more delicate than those of *S. noctiflora*.

The Night-flowering Catchfly can be described as a cornfield weed – and it is also an annual – so that if the field is sprayed or the corn is cut before the seeds have been shed, no plants are to be expected the following year.

It is said to favour sandy soil and is not mentioned in Lousley's *Wild Flowers of Chalk and Limestone*, but, in Sussex and Hampshire at least, it seems to show a preference for the chalk. Both the stem and the calyx carry the true catchfly stickiness so the name given to it in some books – Night flowering Campion – should perhaps be allowed to lapse.

IIIg. *Teucrium botrys* Cut-leaved Germander

There is a certain formality and delicacy of style about this little plant that is all too seldom seen in the wild. Viewed against the bare chalk on which it loves to grow, its structure suggests Louis XV jewelled filigree. The impression is strengthened in as much as the leaves and flowers are all directed to the same side of the stem as might be the ornamentation of a brooch. As with all Germanders, the flowers consist of a single lip, hanging downwards, like a hound's tongue. In colour they have been described as pinkish purple, but the specimens which I have seen on a railway spoil-heap in Hampshire are nearer to a soft fig red.

Thirty or so years ago, plants were to be found on chalk grassland and bare patches of ploughed fields in at least five southern counties – from Gloucestershire, through Wiltshire, Hampshire and Surrey to Kent. But it seldom manages to hold its own against the ranker kinds of vegetation and is at permanent risk from the plough and the earth mover, though it has been known to reappear when fields that have been ploughed are allowed to lie fallow.

In Surrey, the best chances of finding Cut-leaved Germander would be in the areas around Box Hill, on bare ground exposed to the sun, for the plant is a native of the Mediterranean and tends to waste away in areas that are overgrown or shaded. In Kent it was known quite recently in a chalkpit at Upper Halling, near Rochester, although here, too, its survival would depend on local conditions.

IIIh. *Cirsium tuberosum* Tuberous Thistle

* 15th July Wiltshire

The Army offers protection for the best-known site of this rare thistle, for it grows on the Wiltshire Downs in the midst of an extensive military training area lying between Westbury and Tilshead; and you tread there at your peril.

It is a handsome plant, growing at times to three feet in height and it is perhaps unfortunate that Linnaeus and those who followed him chose to distinguish it from other thistles, by the feature that field botanists, we hope, are least likely to see – namely, the roots. These are, predictably, tuberous but are also distinguishable from the roots of some other thistles such as *C. heterophyllum*, the Melancholy Thistle, and *C. dissectum*, the Marsh Plume Thistle, by the fact that there are no stolons – short auxiliary surface roots – in addition to its main spindle-shaped tuberous anchor.

Above ground, however, there is much to admire. The leaves at the base of the stem are deeply and widely cut and green on both sides, and the involucre from which florets arise is globular if not completely globe-shaped. The flowers, usually solitary, are similar in colour to those of a much commoner species, the ordinary Black Knapweed. But, taken in all, the true plant is unmistakable.

It has, however, been known to hybridise with other species of thistle including even *Cirsium acaulon*, the Dwarf Thistle, which sits stemless on the turf ready to startle any unwary picnicker who sits down on it.

The Tuberous Thistle has been reported from Cambridgeshire and

Gloucestershire and Glamorgan. But Wiltshire remains its favourite county.

IIIi. *Teucrium chamaedrys* Wall Germander

* 25th July East Sussex

This modest member of the Labiatae has acquired its 'English' name by virtue of circumstantial evidence. It has been grown as an aromatic under-shrub in gardens ever since the days of Good Queen Bess and is still offered to gardeners by the prestigious seedsmen, Thompson and Morgan of Ipswich. It is also most often to be found on walls – as for example on the ramparts of Camber Castle near Rye in Sussex – where it grows relieved from the competition of other more thrusting species. And so it came to be assumed that any wall Germander had escaped from a nearby garden and was therefore an introduced plant, though in places naturalised, where, being perennial, it had become permanently established. This impression was strengthened by the fact that, in its native Mediterranean habitats, 'Wall Germander' grows on grassy banks, waste ground, etc. rather than on walls.

However, some forty years ago 'Wall Germander' was seen to be established in a small patch on downs to the east of Cuckmere Haven in East Sussex, in a similar habitat to that in which it grows on the opposite side of the Channel in France. The Cuckmere Haven slope is unsheltered and west facing; the turf is short – courtesy of the prevailing wind and some rabbits – and, in consequence, the plants are so small that once found one needs to get a compass fix on a row of posts straddling the down, to be sure of re-finding them. The flowers, dark-rose, are somewhat reminiscent of other labiates which could confuse the questing botanist.

The currently accepted theory is that, since the plants are so much smaller than the garden escape, and since they are growing in what could be their natural habitat, the Cuckmere specimens are in fact native plants and have not been introduced by human agency.

IIIj. *Orchis ustulata* Burnt Orchid

* 29th May Sussex

This elegant little orchid is a miniature of *Orchis purpurea*, the Lady Orchid, and its flowers carry a similar hood or helmet formed over it by the sepals and petals. These hoods when not yet opened at the top of the spike are maroon-black, giving the plant its 'burnt' appearance which gradually vanishes as the flowers develop. In full bloom the sepals fade to almost white as the flowers go over; the flower spike, which was originally more or less cone-shaped, ultimately takes the form of a cylinder.

The lip of the flower is white, three-lobed and dotted with patches of purplish hairs. The central lobe is divided at its base into two blunt wings, sometimes feathered at the edges.

The plant is heavily dependent on fungi during the early stages of development and examination of the rhizome has shown that it can develop underground for thirteen to fifteen years before it is discarded as tubers leading to a shoot and a flower-spike are formed.

From this it follows that the ground on which the plant grows must remain undisturbed during that period if flowers are to appear, and the survival of this orchid would appear to depend on the uninterrupted maintenance of the short-grassed chalkland on which it likes to grow.

Turf which was last ploughed even as long as half a century ago will almost certainly have lost some of the more delicate and thus rarer plants for good. Attempts to 'improve' it with nitrates merely allow the coarser grasses to flourish at the expense of the more delicate downland plants.

On the other hand, if the downs are left to themselves the grassland will turn to scrub which has to be eliminated by burning, cutting or grazing. Some of the coarser grasses can profitably be burnt off in the winter months, but if Chalk False Brome, known as Tor Grass is present, this course should not be followed for burning helps to germinate the seeds and merely worsens the situation. Grazing suits the orchid best since being only a few inches high it cannot compete in long grass.

Sheep which are independent enough to leave the flock and graze among the scrub (yet not so independent that they will jump a fence) are preferred for grazing, especially if they are short-fleeced and thus unlikely to become entangled in brambles, and are hardy enough to overwinter. A useful pamphlet published by the Nature Conservancy Council recommends Beulah sheep from Wales as a useful grazing breed that does not reject the coarser grasses.

Cattle are equally eclectic in their tastes and better at dealing with shrubs, but they break the turf in areas where they congregate for food and water and thus allow new weeds to invade – particularly ragwort. Downland can be grazed either intensively, one area at a time in rotation, or lightly grazed overall on a ratio of one or two sheep to the acre.

Even with all this care, however, the number of flowers of Burnt Orchid varies alarmingly from one year to another. A well-drained slope is preferred but the orchid will stand up to cold in Britain in Northumberland and Cumberland and on the continent, even in Siberia. The Burnt Orchid normally flowers in May or June, but, in parts of Sussex a second flowering of different plants on different sites occurs in July.

IIIk. *Scandix pecten-veneris* Venus's Comb or Shepherd's Needle

* 21st May East Sussex

Those who hold Venus's Comb to be a rather high-faluting name for this humble plant may prefer the equally fanciful alternative, Shepherd's Needle, as being a more suitable description. However, shepherds and goddesses alike are likely to be out of luck in an emergency, for this delicate member of the Carrot family has become increasingly hard to find. It is one of the cornfield weeds which has suffered from the attentions of farmers in their campaign against other weeds far more noxious than this one.

When searching for this plant, one is struck at once by the delicate frizzy appearance of the leaves and the simple character of the 'umbrella' carrying the flowers, which consist of but two or perhaps even only one ray. The flowers, as in most of the Umbelliferae, are white, and one sometimes notices that the outer petals of the flowers in each group are larger than the inner petals – a peculiarity seen on other flowers such as the Candytuft or the Hydrangea.

I first came across *Scandix* one 5th May when hurrying over a ploughed field near the sea in search of the rare Mousetail. With the wind in my eyes it looked just like an absurdly premature Yarrow. But the point of this story is that the year was 1975, when everything seemed to be three weeks ahead of the calendar. The plant was exposed only because the particular farmer who cultivated the field had evidently sown his crop of rape at the conventional time so that the cabbage leaves which in other years would have concealed the Scandix had not yet appeared. Since then I have never succeeded in rediscovering the plant in this particular field, though it may perhaps still flourish unseen both there and in many other fields. The specimens in the photograph were taken on a grassy bank near the outside of Roedean School near Brighton.

Undoubtedly the plant has become scarcer in recent years and it is little consolation to read that it can be found in central Russia and eastwards to the western Himalyas and that it has been introduced in Chile and New Zealand.

IIIl. *Centaurea cyanus* Cornflower

By studying pollen grains which have been preserved in perfect condition beneath layers of peat, pollen experts – or palynologists as they are sometimes called – have satisfied themselves that the Cornflower was flourishing in Britain at a time when the last of the Ice Age glaciations was at its most severe, that is earlier than 10,000 years B.C. On this basis it is deemed to be one of Britain's native plants. However, in later times, as men in settled communities took to growing and sowing wheat, and to buying seedcorn from one another, the seeds of wheat and Cornflower often travelled together, so that it would have been difficult to be sure whether any particular plant was of native stock or had been introduced.

In any case, the modern methods used for cleaning seed (not to speak of spraying and stubble-burning) have turned the Cornflower, whether introduced or native, into a rarity, though it occurs from time to time in the neighbourhood of nusery-gardens as an escape, or as a casual derived from pet-seed, or from the grain used to feed chickens or turkeys.

One thinks of Cornflowers as being bright blue, but this is so only of the florets near the edge of the flower. Those nearer the centre are reddish purple.

Unfortunately it is an annual plant and when found cannot be depended upon to reappear in the same spot the following year. It is also likely, if discovered, to be picked so that its chances of making seed are slim.

IIIm. *Hyoscyamus niger* Henbane

Henbane is one of those plants which may have been unintentionally disregarded by the conservationists because it occurs from time to time in many different areas. But it is extremely local and undependable in its appearances and there is evidence that it may be becoming rarer.

In general it seems to favour disturbed ground near the sea – either on sandy soil or chalk, and is generally regarded as being a native plant in the south and south-east of England. Yet the *Atlas of the Kent Flora* describes it as 'rather local and somewhat erratic'. In Sussex the omens are even less favourable. In 1937 when the *Flora of Sussex* edited by Lieut. Colonel A.H. Wolley-Dod was first published, Henbane was said to be 'Native on chalk Downs, where it is frequent, more rarely and often a casual elsewhere, except near the coast.' Indeed at that time it was so often to be seen on chalk that Wolley-Dod considered it unnecessary to list the chalk localities. Instead, only the stations not on chalk were given – more than two dozen of them stretching right across the county from Chichester in the west and Pulborough to the north, to Winchelsea in the east. But in the *Sussex Plant Atlas* of 1980 we read that it is to be found 'On disturbed ground, usually near the sea. Occasional on the chalk and in sandy places around Rye and Camber' (in the extreme east of the county). There seems to be no explanation of why during the last forty years it has become 'occasional on chalk' instead of 'frequent'. True, some downland has since been lost to the plough, but, since the plant thrives on disturbed ground, and, since the seeds are known to be capable of remaining dormant for years, ploughing should not have made much difference. Spraying would account for its disappearance from some hedgerows and chalky banks, but this would not apply to the large areas of Downs which have still survived.

On the Downs themselves the partial disappearance of rabbits may have been a contributory cause of Henbane's decline. In the past Henbane was often to be found flourishing in rabbit warrens on the chalk, since they offered suitably disturbed ground. Rabbits, it appears, were warned against eating the leaves by the strong smell of the plant. The decline in the rabbit population may have restricted the number and size of the available sites. Henbane is poisonous to other animals besides hens, and when it occurs in farms it is at risk of being cut down before it can seed.

Certainly the plant is striking enough to warrant the trouble of a lengthy search for it. Often it grows to 76 cms (2 ft 6 ins), with a stalk wrapped in soft grey-green toothed leaves covered with soft hairs bearing glands. The flowers are pallid yellow, veined purple with purple anthers. It is usually biennial but the leaves are sufficiently distinctive for the plant to be recognised during its first season of growth.

It was formerly cultivated and prescribed as a pain-killer, particularly to relieve toothache. According to Gerard the seed was 'used by Mountibank tooth-drawers which run about the country, to cause worms to come forth of the teeth, by burning it in a chafing dish of coles, the party holding his mouth over the fume thereof: but some crafty companions to gain money convey small lute-strings into the water, persuading the patient that those small creepers came out of his mouth or other parts which he intended to ease'.

IIIn. *Melampyrum arvense* Field Cow-wheat

* 10th July Isle of Wight

This is one of the showiest, yet least often seen of flowers. It can grow to nearly 61 cms (2 ft), its height depending on that of the surrounding vegetation and on its own vigour. Its tufts of brilliant purple bracts with long slender teeth flare out against the landscape almost as distinctively as lamps on an airport runway. Beneath the topping of these bracts, the flowers peer out on all sides, club-shaped tubes with mouths half-ajar. Each is daubed with purple on the upper lip and with an orange spot on each cheek and on the chin.

It is an annual, and in the six sites where it was last reported to exist, it varies considerably from year to year in numbers. There are but two seeds to each flower.

The botanical family to which the Cow-wheats belong is one of the most interesting of any, including as it does the Speedwells, the Mulleins, the Snapdragons and the Foxgloves. Some species of this family, including this purple Cow-wheat, are semi-parasitic. That is, although they are capable of taking nourishment through their own roots, they will, if given the opportunity, attach themselves to the roots of members of the grass family and live off them.

Melampyrum arvense is usually associated with cornfields and the adjective 'arvense', chosen for it by Linnaeus to describe it, refers specifically to arable land used for growing wheat as distinct from grass meadows. Farmers in the Isle of Wight and perhaps elsewhere too regarded it with special disfavour, calling it the Poverty Weed because, when milled with corn it discoloured the flour and detracted from its value. It is, therefore, not hard to see why it has become so rare, not only in Britain but elsewhere in Europe. Stricter controls now prevent it being transported from one place to another with seed corn. Stubble-burning, a practice followed by farmers partly in order to get rid of weeds, is, of course, another of the hazards facing this plant.

But there is also evidence that *Melampyrum arvense* is not confined to the cornfield – one widely travelled expert tells me none of the British sites are in cornfields but that the plant is happy to grow on banks, verges and in hedgerows. Certainly the plant shown here was one of a group of a dozen on a bank on the border of a field that may once have grown wheat, but which was now supporting tall grasses and here and there a few plants of Field Forget-Me-Not. A second site in the Isle of Wight is also a very steep grassy bank facing south but with some scrub shelter. This kind of habitat, however, could be at risk from crop-spraying, as well as from stubble-burning, and equally from the loss of hedges which are uprooted in order to allow tractors to plough a longer furrow. One of its safer haunts is on bank by a disused pit within a large brickworks – though it may originally have grown on arable land nearby.

The casual manner in which the plant behaves makes it difficult to afford it much protection, and the rate at which the number of sites has decreased raises apprehensions as to its future.

Attempts to establish wild plants of *Melampyrum arvense* in safer conditions are rarely successful – partly because the supply of seed is so limited and partly because the new site may be lacking in suitable host plants for the seedlings. The flowers are normally pollinated by bumble-bees attracted to the nectar at the base of the corolla. But fertile seed can also be obtained by self-pollination.

The plant flowers almost throughout the summer and one is entitled to hope that, if present, it will not be overlooked by those seeking to protect it.

IIIo. *Melampyrum cristatum* Crested Cow-wheat

* 15th July Bedfordshire

Though more unusual in some ways than *Melampyrum arvense*, this species is not quite so rare. It has been recorded from Wiltshire, Hampshire and Essex and as far north as Nottingham and Lincoln, whereas *M. arvense*, apart from the Isle of Wight, is limited to Bedfordshire and Essex.

The Crested Cow-wheat favours copses, the edges of woods, verges and hedgerows rather than the cornfields as such. In this case, too, it is the bracts rather than the flowers themselves which are distinctive. They are bright rose in colour and reminiscent of a rooster's comb. But when the flowers are fully developed these toothed bracts are seen to be overlapping one above the other like tiles. The architectural flavour of the plant is heightened by the fact that the blooms are disposed in a four-sided spike. The tube-like blooms are variable in colour, some being completely suffused with rosy purple whereas some are but faintly tinted. The lower lips of the flowers are slashed with deep yellow. But the plant lacks the 'shocking purple' top-knot which makes its near relative so impressive.

John Ray in his *Synopsis Methodica Stirpium Britannicarum*, published in 1724, records it as 'In sylvis Cantabrigientibus & Bedfordiensibus copiose' and adds: 'In Walton-field, near Wakefield, amongst the corn; Dr Richardson'. Dr Richard Richardson (1663–1741) was one of 45 contributors to the *Synopsis*, seventeen of whom were physicians. Among them were Sir Hans Sloane, whose collections provided the nucleus for the foundation of the British Museum, and Sir Robert Sibbald after whom the little yellow-flowered wild stawberry-like plant *Sibbaldia procumbens* was named.

Both *M. arvense* and *M. cristatum* are calcicoles – that is, chalk-lovers. But *M. cristatum* also particularly favours those areas containing what is known as boulder clay. Its origins stretch back to the Ice Ages. These glaciers which planed the surface of the land, drove soil and rocks before them, grinding away the chalk surface and compressing it into boulders –mechanically formed rocks containing a reservoir of chalk. Boulder clay is found in Cambridgeshire to the south and east of Cambridge in such places as Hardwick Wood, Caxton, Eversden and Long Stowe, and in Bedfordshire in such places as Hatley, to the east of the county. The photograph shown, however, was taken in a site further north and in flat, open countryside.

Probably the best-known example of a plant that seems to thrive on boulder clay and nothing else is the Oxlip, *Primula elatior*, a species which resembles a Cowslip but with more open Primrose-like flowers, stalked in an umbel (little umbrella) and drooping to one side. Those with a lens will note a feature which distinguishes it from the much more common hybrid between the Cowslip and a Primrose. The corolla tube is without folds in the throat. Seeing how many Primroses are heedlessly picked in the springtime, one might have imagined that this was a plant at risk. But it is well-protected by the Suffolk Trust for Nature Conservation who have a number of sites open to members.

Melampyrum cristatum is one of those woodland plants that responds to coppicing. That is when it springs up in quantity after the nearby underwood has been cut down but gradually dies down again as other more rampant plants crowd round about. This suggests that the seeds – four to each flower – can lie dormant in the soil for several seasons. If *Melampyrum arvense* has the same kind of dormancy, this may help to explain why its numbers vary so sharply from year to year in its special sites.

IIIp. *Cuscuta europaea* Large Dodder

This species with golf-ball heads of pink flowers is a parasite which attaches itself with suckers to a host plant – usually the Stinging Nettle – winding itself round the stem and indulging itself in spaghetti festoons of utter confusion. The stems of the Large Dodder are up to 1 mm thick – about ten times that of the Common Dodder most often to be seen on heather or gorse. Moreover, the stems of the Large Dodder, though sometimes reddish, are more often the colour of tripe, which has earned them, in Sussex at any rate, the nickname Devil's Guts.

The Large Dodder is a rare and decreasing plant, although for no readily apparent reason. It is often to be found on the banks of streams such as the Mole in Surrey and other tributaries of the Thames, and it has been suggested that the seeds are sometimes waterborne. Possibly water pollution has reduced the chances of survival. Moreover, river banks, particularly in southern Britain, are at risk from developers; farmers often try to clear nettles from their water-meadows if cattle are to feed there. The plant also occurs in hedgerows, where it is equally at risk from large-scale farming practice.

Both the Large Dodder and its smaller cousin are European, so there seems to be no particular reason why the larger plant should be specially designated as such – particularly since it occurs as far afield as Tibet. The vernacular name 'Dodder' seems to have travelled with the plant from Germany centuries ago but has no known derivation.

IIIq. *Salvia pratensis* Meadow Clary

* 19th June Gloucestershire

The aromatic scent was enough to convince the apothecaries of the past (and their patients) that a touch of Clary was the surest way of clearing the eyes. The approved technique was to place a seed in each eye and wait until they – the seeds – dropped out.

But, for better or worse, our own Meadow Clary does not take readily to the flower-bed and botanical gardeners therefore preferred to turn to such varieties as Cretan Sage, of which the smell was considered 'more vehement'. But for this *S. pratensis* might still have been a common plant.

As it is, it grows in Britain in small colonies on perhaps no more than a dozen sites – most of which are on private land which means that the plants are not always under the protection of a Nature Conservation Trust.

The plant is also at particular risk in as much as it prefers chalk or limestone grassland that has not been 'improved'. Research by the Nature Conservancy Council has shown that between 1966 and 1980, taking samples from sixteen critical counties, around 70 per cent of this type of habitat had been lost for good, and the remainder was under threat. Most went to the plough; some was overgrown by scrub and elsewhere plants were out-manured, crop-sprayed or eaten by rabbits.

Those that still remain are magnificent – often a yard or more high standing out above the summer herbage – a glorious patch of blue with individual flowers of up to 2.5 cms, their upper lips compressed at the sides to give the characteristic arched 'roof' to the flower. The plant shown in the photograph was in a perfect summer meadow on a gentle slope with a village above and a country lane beneath. The plants are just out of sight of the village street and the Trust concerned have – or had when I visited it with their permission – given no clues as to the treasures contained in the field under their protection.

The oldest known site for *Salvia pratensis* is in the area of Wychwood forest in Oxfordshire, where it has been known for at least two hundred years. There, at least, it must be regarded as native, even though, in some areas it could conceivably have been introduced with grass seed.

In continental Europe *S. pratensis* has been recorded from Scandinavia but it prefers a warmer climate and its range extends from central Spain and Morocco in the west to the Crimea.

While no one will mistake the *S. pratensis* once found, there is a lesser plant – far more common – which could delude the unwary into thinking that they had stumbled on the real thing. This lesser plant is *Salvia horminoides*, Wild Clary. The flowers of this are smaller – only up to 15 mm in length and shorter than the calyx, which is clothed, particularly at the base, with long white hairs – not present in *Salvia pratensis*. The Wild Clary also carries two white spots on the lower lip of the flower, and the leaves, near the roots, are not long and narrow like those of *S. pratensis* and are seldom twice as long as they are broad. The aromatic scent is also less in evidence.

IIIr. *Orchis purpurea* Lady Orchid

* 14th May East Kent

This beautiful orchid with polka-dot lip and dark purple helmet is rarely, if ever, seen outside Kent. In that county there seem to be two distinct forms. Those that flourish west of the river Stour, which runs through Ashford and Canterbury to Sandwich, are shorter, more densely flowered and more heavily spotted than those growing to the east of this line. The plant prefers a chalk soil and some, though not too much, shade. Indeed, experience has shown that fresh shows of plants are obtained if new areas of woodland are cleared as the old ones become overshadowed by the broad-leaf canopy.

It is often to be found with other woodland chalk orchids such as *Ophyris insectifera*, the Fly Orchid, *Orchis mascula*, Early Purple Orchid, and *Listerata ovata*, Common Twayblade.

Anyone wanting to see these orchids at their best should get in touch with – and contribute to the expenses of – the Kent Trust for Nature Conservation, 125 High Street, Rainham, Kent, ME8 8AN which does so much to safeguard the future of these and other treasures.

IIIs. *Lobelia urens* Heath Lobelia

A true mid-blue flower this – better in colour than the cornflower – and equally rare – being confined to about ten localities – Cornwall, Devon, Dorset, Hampshire and Sussex. The edges of damp, acid woods are favoured habitats. In Sussex it reappears and disappears in a muddy copse at Flimwell close to the borders of Kent. When this wood has been coppiced there is a good chance of seeing the plant in quantity, but when the trees shut out the light, no Lobelia.

Heath Lobelia is usually accepted as one of our native plants, although the Sussex locality in which it persists was not discovered until 1920, which has led some people to assume that it must have been introduced there.

The genus Lobelia was named, after his death, for Mathias de l'Obel (1538–1616) whose famous book *Stirpium Adversaria Nova* – A New Notebook of Plants – was dedicated to Queen Elizabeth I. Mathias de l'Obel was for a time Superintendent of Lord Zouche's garden at Hackney and had a hand in producing, and later improving, John Gerard's *Herbal*.

IIIt. *Gagea lutea* Yellow Star-of-Bethlehem

* 12th April Gloucestershire

This attractive plant claims one of the earlier dates in the botanist's calendar and is often in flower in March. It is a member of the Lily tribe, and, of course, a cousin of the still earlier flowering *Gagea bohemica*. It favours open copses where the soil is non-acidic and 'basic', i.e. rich, full of minerals. It is often in company with *Endymion non-scriptus*, the Wild Hyacinth alias 'Bluebell' (though the latter can flourish in acid woods too).

Although not included in the *British Red Data Book*, the Yellow Star-of-Bethlehem would seem to be a vanishing species and is absent from counties where it was once well known. Its disappearance may be due to tree-felling, but sometimes its absence may be temporary as it frequently fails to flower for years at a time on sites where it is known to have become established. During the 'off' years all that appears is a single leaf without a hint of the accompanying flower stalk and bracts. The leaves are very similar to those of the Bluebell and are distinguishable from them by two ridges running down the undersides of the leaves. Only the dedicated botanist will wish to satisfy himself by looking at the underside of every leaf in a Bluebell wood. But it sometimes happens that the leaf of *Gagea lutea* is contracted at the tip into a hood – a feature which is not shared by the Bluebell – and this can be taken as a sign that the plant, though not flowering, is surviving.

The flowers, when they appear, are pale yellow, of a colour warmer than lemon but without the glow of the palest buttercup. There may be as many as five flowers on a single plant, with the stalks appearing to have spread, umbrella-fashion, from a single point. Like the Wild Tulip they carry a green stripe on the reverse of the petals.

IIIu. *Silene otites* Spanish Catchfly

* 6th July Norfolk

Once upon a time this unusual plant was known to botanists as *Sesamoides Salamanticum magnum* – the Greater Spanish Catchfly. John Gerard gives no better explanation for the Spanish connection than that its relative, the Lesser Sesamoides, grows in a stony soil on the hills near Salamanca. He called the plant the Great Bastard Woade. Its status as a native of Britain seems to have been in doubt for some time. In Ray's *Catalogue of Plants Growing near Cambridge*, it was said to be 'near the gravel pits as you go to the nearest windmill on the north side of Newmarket town'. But a note beneath added that the plant was suspected of being an alien. This caveat is not however repeated in Ray's later work *Synopsis Stirpium Britannicarum* where it was said to grow: 'In and about the gravel pits on the north side of Newmarket Town: also by the waysides all along from Barton Mills to Thetford in Norfolk.'

Today it is to be found in perhaps thirty sites in Norfolk, West Suffolk and Cambridgeshire, and must be regarded as being at considerable risk since the area of its foothold in the Suffolk Breckland suffered a cut of 50 per cent during the thirty years which followed the end of the Second World War.

John Gerard, being a gardener, had little time for the Spanish Catchfly and declared that it 'hath very long leaves and many, slender towards the stalk, and broader by degrees towards the end, placed confusedly upon a stiff, thick stalk on the top of whereof grow little foolish or idle white flowers: which, being past, there follow small seeds like unto Canarie seed that birds are fed withall.' The petals in fact are not white but pale yellowish green, shaped like a fly's wing and the branched stems and prominent anthers give the plant a 'fussy' appearance.

Though unlike the other Catchflies to look at, the Spanish Catchfly has some of their characteristics. Its stem is sticky, its seeds paddle-shaped, and, like *Silene noctiflora*, the Night-flowering Catchfly, it is scented at night.

Its preference for East Anglia must be because the rainfall there is the lowest in the country. It is essentially a plant of the Steppes and flourishes over a wide area of western Asia. July is the right month to look for it.

— IV —

Plants close to the Brink

* (See facing pp. 130 and 131 for illustrations)

IVa. *Matthiola incana* Hoary Stock

* 8th June Sussex

Plants that live on a billet close to the tide-line are at risk on five different fronts. There are the cliffs themselves which may fall down, and then there is the turf close to the cliff top, sometimes equally precarious. On a lower plane there are the saltings and perhaps close to them the sand dunes, both liable to be grazed over or even tipped on. Then there is the turf and also waste places of a sandy character, the haunt of bathers and ice-cream vendors, and finally there is the shingle on which some of the rarest plants have nevertheless succeeded in establishing themselves. Probably the most irreparable damage occurs from man's efforts to protect the coast-line from erosion by the sea. Walls are built; tracks are established to carry cement-mixers and bricks; loads of spoil are dumped here and there; the earth is disturbed and brings forth docks, thistles and brambles, and, though some land may be saved from the sea, the plant is lost. In areas that have not already been built up, rabbits can be a menace. And in holiday areas it is difficult to cordon off an area effectively as a reserve – and to keep it free of cans, cartons and other cast-offs.

Two very different rarities are attracted to the cliffs. In the first rank and at a high risk, one would put *Matthiola incana*, the Hoary Stock. It is so named because its mop of greyish green leaves are frosted with white hairs. The petals are sometimes white too – nearly all in the case of plants growing above the Brighton Yacht Marina. But those with petals of Tyrian purple are the more handsome. It is 'Doubtfully native' according to Clapham, Tutin and Warburg's *Flora*. Be that as it may, it would have been an adventurous gardener to have introduced it to the sites on which it now flowers. Apart from the chalk cliffs to the east of Brighton, it is now to be found in similar situations on the Isle of Wight. There it has been recorded on the cliffs between Ventnor and St

Lawrence in patches, and on the chalk cliffs of Afton Down towards the western extremity of the island. It was first observed in the middle of the nineteenth century by Dr Bromfield. But the Sussex records go back much further – to 1808.

Elsewhere there are records of it being an introduced plant which is hardly to be wondered at, since it is not only fair to see, but delightfully scented. It was reportedly introduced into the Channel Islands during the reign of Good Queen Bess.

At large, *Matthiola incana* is often a perennial but the cultivated variety, known as Ten Weeks Stock, is an annual as is the even more energetic Seven Weeks Stock. Thus there is no need for plant lovers to put their own lives or that of the wild plants at risk.

The wild stocks should be at their best during the fourth week of May, but, to appreciate them, take with you a small pair of binoculars.

The chief danger to the wild plant comes not from enthusiastic gardeners but from the fragile nature of the chalk cliffs on which they love to grow. A glance along the foot of the cliff will usually show tell-tale evidence in the form of talus – the geologist's term for débris – which accumulates there from falls of the chalky rock. Often the plant is brought down too. There are, naturally, cliffs and cliffs. Some are of hard granite or red sandstone which stands up to almost anything that the weather can throw against it. Some limestone rocks, as at Stackpole Head in Pembrokeshire, are also weather-resistant as is the chalk of Flamborough Head in Yorkshire.

Other forms of limestone are easily worn away by as much as a metre a year. And then there are even softer, unstable cliffs based on soft sands or clay which give way, periodically giving rise to new landslips of which the coast to the west of Lyme Regis is an outstanding example.

Many of the cliffs – particularly those on the west coasts of England and Scotland – have been declared as Sites of Special Scientific Interest so that no unauthorised quarrying or other forms of disturbance may take place.

Some of the chalk cliffs on which *Matthiola incana* grows have some protection in as much as a fence has been installed to discourage ramblers from approaching the edge. In places the fence seems surprisingly close to the cliff-top; but at others where the cliff face has resisted the weather, there is plenty of space between the cliff-face and the coastal footpath and on this verge a number of plants succeed in maintaining their rights.

Since the Stock favours the upper part of the cliff, one may assume that, although it tolerates the salt of sea-spray, it is not so tolerant as for example Samphire, or the so-called Hottentot Fig which covers large parts of the low cliffs in north Devon. The Stock does not appear to need either the manure from the droppings of sea birds nor yet flushes of water trickling from springs in the cliff-face.

Matthiola incana was beloved of gardeners as far back as the sixteenth century when it was known as Stock-gilliflower in order to distinguish it from those other scented July flowers, the Pinks, which had no stock or woody base to their stems. In most cases, only plants with double flowers were grown in the show gardens, and it was recommended that the seed be mixed with those of the radish as a deterrent to the flea beetle. Of these plants John Gerard, the herbalist, wrote 'They are referred unto the Wall-floure, although in vertue much inferiour, yet they are not used in Physicke, escept among certaine Empericks and Quacksalvers, about love and lust matters, which for modestie I omit.'

The genus Matthiola was named after Pierandrea Mattioli (1501–1577) born in Siena and brought up in Venice where his father was a doctor. His parents intended that he should take up Law, but he preferred medicine – in particular that branch which had to do with medicinal herbs. After practising in Siena and Rome, he was appointed physician to the Archduke Ferdinand and later to his brother, the Emperor Maximilian II. His fame as a botanist rests on his *Commentary on the works of Dioscorides* which did so much to reconcile the plants described by the Greek master with those actually to be found 1,400 years later. Mattioli's work, published in Venice in 1544, contained about 500 true-to-life illustrations in woodcut, and ran to forty editions.

IVb. *Aster linosyris* Goldilocks Aster

* 20th September Somerset

One would not have thought at first sight that this would be a seaside plant, for it looks too slender and willowy to stand up to the blustering winds of the coastline. Yet it clings to the sea-shore, particularly in the west, though nowhere in large quantities.

There is a small stand on Berry Head in south Devon, and here it shows its strong preference for chalk or limestone soil, disappearing abruptly at the spot where the basic soil peters out. In Somerset there is another small colony on an outcrop of suitable rock – again near the sea, and on part of a turfy escarpment above it, where it is in danger of being overrun by brambles.

On the Gower peninsula in Glamorgan, botanists, among them David McClintock and the late Ted Lousley – who have checked the likely rock outcrops – have found at least two small groups in the western part of the peninsula. There are two more colonies in Caernarvonshire on Great Ormes Head, which botanists visit for many other rare plants, and Pembrokeshire has its own little assembly at Castlemartin, south-west of Pembroke, in an area so remote that the Army uses it for manoeuvres. And then there is a station in north Lancashire on that finger pointing south into Morecambe Bay known as Humphrey Head. There it should be safe enough, for it is cliff country and Goldilocks grows in a more or less inaccessible position – again with other rarities such as Spiked Speedwell and White Rockrose.

The plant is notable for the profusion of green almost hairline leaves spiralling up the stem. If they were but a little broader, they and the golden flowers above would remind the onlooker perhaps of that garden favourite, Golden Rod. In fact, however, botanists describe the leaves as linear or line-like, and the impression of a fringe of green hair is so vivid that at one time the plant carried the botanical name *Crinitaria* from the Latin word for hair.

The flowers are arranged at the top of the flowering stem in a corymb, that is a group of flowers of which those in the centre are on shorter stems than those at the edge of the circle – with the result that all are more or less level.

The individual blossoms are composed of golden disc-florets without the external rays that decorate so many other composite plants such as, for instance, the sunflower. But they make a good show.

It was a surprise to learn that this rarity has in the past been recorded in other counties further to the east – including Sussex. For example, in Wolley-Dod's *Flora of Sussex* (1937), it is described as 'a casual alien. Sea-shores; very rare'.

The first record was in *The New Botanist's Guide* (published 1835–7), and records a plant growing 'between Shoreham and Brighton – W.C. Trevelyan fide H.C. Watson'. This was presumably Hewett Cottrell Watson (1804–1881) famous among botanists for a number of other standard works including the classic *Topographical Botany*. Several plants were also said to have been found in 1926 at West Wittering almost at the western extremity of the county by a Mr S. Morris. However, the seeds would certainly have had to travel several hundreds of miles from one source or another to reach Sussex, because it is essentially a central European and even Mediterranean plant. Equally mysterious is the fact that the locations where the plant persists are so widely separated from each other, and that the plant never seems to spread.

One explanation that finds favour among the experts is that the sites in which it still survives are ones that escaped being covered with sheets of ice during the Ice Ages. In this its history would have differed from that of many other plants which retreated to the south each time the ice advanced, and advanced northwards again as the ice thawed out. The Goldilocks Aster had no need to exercise so much adaptability – and equally had no capacity to expand from its sheltered position.

And can we deduce anything from the fact that, like its nearest relative, Sea Aster, it flowers so late in the year – typically during the third week of September? Is it waiting for a new warm age that never comes?

IVc. *Ophrys sphegodes* Early Spider Orchid

* 5th May East Sussex

For the next batch of seaside rarities, we must look at the turf on top of the cliff, a little way from the edge. What grows there depends on the agricultural practice of the area. Clearly if arable farming is practised under modern conditions, the plough will come close to the cliff-top and the area in which plants can grow at random is greatly restricted. If in addition neither sheep nor rabbits graze the margin of the cliffs, then in time scrub will take over, and as at Humphrey Head in Lancashire, there will be trees growing vertically on the face of the cliff itself. If, on the other hand, rabbits are at hand, or even if the winds are sufficiently strong – as for instance at Beachy Head in Sussex or Berry Head in south Devon or Nwnt Cliff to the north of Cardigan, then smaller plants such as Spring Squill can prosper. The pink miniature Scottish Primrose that grows on Cape Wrath and the smaller forms of Centaury depend on such conditions.

But cliff-tops are more vulnerable than cliff-faces. Cliff-head paths encourage visitors to tramp to and fro, destroying the vegetation and causing erosion of the cliff itself. Picnic centres, view-points, car parks and even caravan parks compound the damage.

One rarity that has so far survived is *Ophrys sphegodes*, the Early Spider Orchid. This attractive orchid appears early in May and is found on chalk downs on parts of the Dorset coast as well as on downs near the sea in Wiltshire, Kent, Sussex, Cambridge, Suffolk and north Wales. The first bloom appears when the plant is but a few inches high, but the spike develops to more than a foot and may have six or more flowers.

The flower is somewhat like that of the Bee Orchid. But whereas the sepals of the Bee Orchid are of a rosy pink, those of the Early Spider are palish green and the lip dull purplish brown, and unlike the Bee Orchid it has no appendage at the base of the lip. The lip is velvety in texture except for an area in the centre, which carries a number of glossy blue-violet markings, often in the shape of an 'H'. In time the colour of the lip fades into yellow and the markings lose their lustre. Occasionally the markings on the lip are in the shape of a horseshoe which helps to justify the *Ophrys* part of its name, which comes from the Greek word for eyebrow. To many, however, it does indeed suggest the body of a large, fat, round Garden Spider.

This is one of the plants which, sad to say, is decreasing. It is now only south of the Thames in about 17 localities, in Sussex, Dorset, Kent and Gloucestershire. Records north of the Thames date mainly from the early part of the nineteenth century, and the decline seems to have begun before the increase in arable cultivation during the Second World War and before the introduction of milk quotas which resulted in the loss of so much downland.

This is an attractive subject for the camera but the photographer is faced with a dilemma. If the shot is taken while the first flower is at its best, the plant will still be short of its full stature – which is up to 30 cm (nearly a foot). If, on the other hand, the picture is not taken till the plant is at full height only some of the blooms will be in perfect condition. In either case, it may help to choose a plant growing on the brow of a slope so that it can be taken against the skyline, though in this case the flower is not always facing in a favourable direction.

The Early Spider Orchid was described in the third edition of Ray's *Synopsis Methodica Stirpium Britannicarum* (1724) in which it was called Humble Bee Satyrion with Green Wings 'growing on dry and gravelly soils in an old Gravel pit in the open field near Great Shelford in Cambridgeshire, and by the roadside near Bartlow'. The note added that according to Dr Richardson it needed chalk soil, flowered in April and was rather frequent on chalk near Northfleet.

IVd. *Orobanche caryophyllacea* Bedstraw Broomrape

* 18th July Kent

We have now left the cliff-top and are searching for some of the rare plants which grow on turf near the sea but nearer to sea level. Perhaps they are able to tolerate a larger pinch of salt than the elevated plants near the cliff-head. One of the rarest of these is the parasitic plant *Orobanche caryophyllacea*, Bedstraw (formerly Clove-scented) Broomrape, which is known only at two sites in Kent, neither far from the sea and, as no other stations are known, one must assume that the plant prefers a maritime environment. Those with a highly developed sense of smell claim to have detected the scent of cloves coming from the flower. I have not enjoyed that experience, but I can say that one should accept with caution the reiterated statement in books that this plant is parasitic on Hedge Bedstraw – a common, white-flowered member of this not very exciting family. The Clove-scented Broomrape to me was clearly more interested in another species of Bedstraw with yellow flowers which grows nearby, namely *Galium verum*, or Lady's Bedstraw, perhaps because it, too, is scented.

Being parasitic, Bedstraw Broomrape has – and needs – no chlorophyll and there is not the slightest suggestion of green. But when first coming into flower the whole plant is suffused with the most delicate roseate pink, and the flowers with their frilly lips suggest a femininity which, alas in time, disappears as the flowers change to reddish brown. Anyone who consults the *Atlas of the Kent Flora* will learn that a good number of plants are to be found on fixed sand dunes in Sandwich Bay. I should perhaps add that the dunes in question are not on public land and the casual visitor is not free to range at large.

Clove-scented Broomrape is rated both in Britain and on the continent of Europe as one of the plants most at risk. This is because the number of sites on which it grows has decreased rapidly because it is an attractive plant, because they are now so few, because they are not remote or inaccessible and because the dunes on which the plants are found cannot easily be turned into a nature reserve.

I notice that there is a tendency to distinguish species members of the Broomrape family according to the plants on which they feed (e.g., Knapweed, Carrot, Yarrow, Thyme, Oxtongue and Thistle). But the rule is not invariably applied to all members of the genus. The most common Broomrape of all and the one most often seen, *Orobanche minor*, feeds on too many different plants for it to be linked with any particular one, and the Greater Broomrape feeds mainly on Broom, so that one would have to call it the Broom Broomrape which would sound ridiculous. So even without taxing everyone's sense of smell, Clove-scented Broomrape seems an acceptable name to have been retained for this very attractive plant.

IVe. *Trifolium stellatum* Starry Clover

* 26th May Shoreham

Starry Clover is an introduced plant which has been here long enough to be treated as 'one of ours'. It appears to have been recorded as far back as the first years of the eighteenth century, in circumstances that suggested it had been introduced and it has invariably occurred in counties bordering on the sea: Suffolk, Essex, Kent, Sussex, Glamorgan, Co. Down, and the Channel Islands.

In Sussex the first mention occurred in a report by William Borrer (1781–1862) which stated 'found July 30th 1804 between Shoreham Harbour and the Sea, growing in great plenty'. A suggestion of how it might have got there was given in *English Botany* edited by J. Boswell Syme (published 1863–1872). There it was said to be 'perfectly naturalized on the ballast along Shoreham Harbour where it has maintained its position since 1804'. In other words, while the Starry Clover may have occurred elsewhere as a casual visitor, at Shoreham, at least, it had found a permanent home.

According to a later report some plants arrived in the ballast of ships returning from Wellington's expedition at the start of his Peninsula campaign. Certainly at large it is a Mediterranean plant.

From the fact that the clover was first established on the area between Shoreham Harbour and the sea, it is clear that the visitor coming from the north who wishes to find it must cross the iron Norfolk bridge from Old Shoreham and enter Shoreham by Sea which lies on the right bank of the river. During the Second World War this became a heavily fortified area with lavish festoons of barbed wire. But when peace came it was redeveloped with long tarmac avenues and drives of houses accommodated to the curves of the river.

Faced with such discouragement it would not have been surprising if the Starry Clover had joined the list of Britain's extinct plants. However, it has not only survived but at times has prospered. For even today there are strips of grass along the avenue, grass on the centre of the roundabouts, waste ground near the river and the occasional unoccupied building site. It is in these locations – for it is an annual and may appear at different spots in succeeding years – that the pilgrim must look for it. At various times it has been recorded on both sides of Shoreham's river, the Adur, and it would not be surprising to find it turning up on the nearby airfield.

The general impression is of a feathery plant because the stem, and the leaves, are covered with fine white hairs and the calyx tube with shaggy white hairs. The strawberry-shaped heads of flowers are massed together at the end of the stalks. As the flowers go over they turn from white to pink, and the calyx opens wide revealing the sharp teeth of a five-pointed star. Some other clovers have a calyx with teeth which turn outwards after flowering but none are so prominent as those of the Starry Clover which are three times as long as the calyx tube itself.

This must have been all that Borrer could have seen when he noticed the plant at the end of July, as the flowers make a good show as early as the last week of May.

A note in the *Flora of Sussex* states that 'the existence of the plant and its exact habitats are well known to many living in the neighbourhood, including such people as taxi-drivers, and it is known to have been exploited for profit, so it may be preserved for this purpose'. *The British Red Data Book for Vascular Plants* takes a less sanguine view, placing it on a par with the Lady's Slipper Orchid on the grounds of rarity and, because it grows in a developed area, is readily accessible, cannot easily be protected and is moderately attractive to gardeners. Time will tell.

IVf. *Matthiola sinuata* Sea Stock

* 14th July North Devon

This plant, in contrast to *Matthiola incana*, is definitely native and regularly biennial.

Though the flowers are of a wishy-washy mauve and are, one would have thought, less attractive than those of the purple form of *Matthiola incana*, the species is thought to be at much greater risk – partly because of the rate of decline and partly because its habitats along the sea-shore are more easily accessible to the general public. In bygone days it grew in Pembrokeshire, Merioneth, Caernarvonshire, Anglesey and in Ireland at two stations, one in County Clare and the other in County Wexford. It is found now only in Glamorgan, Devon and the Channel Islands.

The Devon site at Saunton Cliffs, mentioned in *Finding Wild Flowers* by R.S.R. Fitter, is the most convenient since it is close to two excellent hotels. Though the cliffs at Saunton provide a useful landmark, the Great Sea Stock, as it is sometimes called, grows at their foot and no exertion is required to get a close-up view.

The stem and leaves of this plant have the same hoary look as *Matthiola incana*, but the leaves are marked with black dots and toothed with a wavy sinuous outline – hence the botanical name. It is scented but only at night. It would seem fairly clear that attempts to remove the plant and grow it in any ordinary garden would surely prove unsuccessful. Fortunately an acceptable alternative is available from seedsmen in the form of *Matthiola bicornis*, Night-scented Stock.

The wild plant has a deep tap root which enables it to draw up moisture from the otherwise inhospitable terrain. A distant cousin of Sea Stock, namely Sea Rocket, also has mauve four-petalled flowers (like those of all the Cabbage family) and can spread without difficulty from dune to dune because its seeds can float and still remain viable in sea water, an accomplishment Sea Stock lacks, and the problem of forming a nature reserve to protect it on any sea-shore popular with holidaymakers is a formidable one.

Matthiola sinuata was also known in the early seventeenth century and appears in the 1633 edition of Gerard's Herbal as *Leucoium marinum purpureum L'obelii*. A note, apparently from the editor, Thomas Johnson, adds: 'The figure of Lobels which here we give you was taken of a dried plant, and therefore the leaves are not exprest so sinuate as they should be.' It appears in the 1759 edition of Linnaeus' *Flora Anglica* as *Cheiranthus sinuatus*, but is missing from the earlier edition of 1754.

Matthiola sinuata grows near the edge of Braunton Burrows, one of our largest sand-dune systems to which we shall be referring later. The first sand dunes would have begun to build up in Britain from the time when the level of the coastline started to stabilise – perhaps some eight thousand years ago, even before Britain became separated from the continent by the waters of the English Channel.

Dunes are formed from sand blown on to land by onshore winds. One extreme example occurred in Scotland during the great storm at the end of the seventeenth century which led to the formation of the Culbin sands. So much sand was blown ashore that, in a few years, the land that had formed part of an agricultural estate was buried beneath the sand and had to be abandoned. The village of Forvie in the Grampian area of Scotland suffered the same fate in 1413 and had to be given up, though it had been occupied for 1,500 years. Today the name survives only in the headland known as Forvie Ness on the seaward side of the sands of Forvie.

Most of the sand forming the dunes is the remains of sediment ground up by the glaciers of the Ice Age. The sand collects round any obstacle, and an artificial sand dune can be created by planting hurdles or palings in the beach. These reduce the speed of the wind and the sand in consequence falls to the ground.

If Marram grass, which grows readily in sand, is planted, the sand becomes anchored, and at this stage the dunes are described as yellow or white from their general coloration. But as the sand firms other shoreline plants such as Sea Bindweed, Prickly Saltwort, Orache, Sea Holly, and Sea Rocket gain a foothold, and mosses and lichens begin to carpet the flat earth between the ridges of Marram. The dunes are then referred to as grey dunes.

In the next stage, depending on the numbers of rabbits present, larger plants such as Docks, Tree Lupins and Sea Buckthorn develop, giving an entirely new look to the landscape. If, however, the life cycle of the dunes is disturbed and the sand loosened by trampling, horse-riding, motor-cycling, picnic fires, and the like, then the wind scours away the sand once more and what is known as a 'blow-out' occurs and the dune has to be re-established all over again.

Marram grass, through which the original stabilisation would have occurred, tends to grow vertically in clumps connected to one another by runners. The result is that the dunes have their highs and lows, with knolls and bluffs and ridges with slacks in which the rainwater collects. Soil at the base of these slacks, having originally been blown in from the sea, contains large amounts of calcium from the shells of countless marine animalcules, so that in time, when a layer of grasses living and dead has been superimposed and some of the salt has been leached out by the rain, you get an area not unlike the chalk slopes of the downs. This helps to account for the abundance on some dunes of orchids, some of which grow almost nowhere else.

None of the dunes are immutable. The feathery Tamarisk shrub with its dense flowered pink racemes alters the character of the dunes or sea-banks on which it grows. It extracts the fresh water from the soil and transpires it into the air, leaving ever-increasing quantities of salt in the ground.

At Ainsdale Dunes near Southport, conifer plantations have lowered the water table, favouring the growth of the local speciality *Epipactis dunensis*, Dune Hellebore.

For some years now, all-out war has been declared wherever Sea Buckthorn, the shrub with silvery foliage and brilliant orange berries, has started to overrun the dunes. It tends to shade out the smaller plants and co-exists with bacteria which increase the nitrogen content of the soil which by no means suits all the most desirable plants.

Braunton Burrows extends along the Devonshire coast for about six kilometres northwards from the estuary of the Taw and Torridge Rivers. It is owned by the Christie (of Glyndebourne fame) Estate Trustees and about two-thirds of the dunes have been declared a national nature reserve and is managed by the Nature Conservancy Trust and voluntary wardens.

For four centuries naturalists have marvelled at the profusion of flowers that carpet the settled portions of the dunes. And even if there were no rarities, the rich masses of colours spread over the ground, filling the air with scent during the summer months, impress even the most casual visitor and are a paradise for the botanist. Brilliant yellows come from the Biting Stonecrop and Evening Primrose, pools of purple from wild thyme, and in places it is almost impossible to find a pathway through the dense stands of Marsh Orchids and Marsh Helleborines. A board-walk across the dunes is provided so that visitors who want to bathe can reach the sea without destabilising the dunes. We shall return to them later.

IVg. *Allium babingtonii* Babington's Leek

* 5th August North Cornwall

Still in dune country we are at the church of St Enodoc in Cornwall. It was once so nearly overwhelmed by sand that, when Arthur Mee wrote about it in 1937, he recalled that in earlier times when the parson and his clerk came to take the service, which they were obliged to do in order to qualify for the tithe revenue, they had to enter the church through the roof. For a time the building was lost altogether, and was not recovered until the last century.

It stands still in the country beloved by John Betjeman, who spent there happy hours that he longed in vain for the rest of his life to recapture, hours commemorated in his poem *Sunday Afternoon Service in St Enodoc Church*. On the way across the golf course to the service, the poet savoured Lady's Finger Vetch, Wild Thyme (and, later, sea-pink and stone-crop), while during the service he could hear the rollers close at hand and notice a bee flying to freedom behind the stained-glass window.

It was this poem, and a memory of Betjeman at Oxford, that took me to the church one August morning. Then, instead of returning across the links, I stumbled along close to the shore-line, and across a small ditch through which a trickle ran down towards the sea. And there on the edge of the rythe was this two-storey leek – *Allium babingtonii* – Babington's Leek. It is not a Scheduled Plant, but rare enough to be included in *The British Red Data Book*.

This is a native plant and would seem to have been properly described first by William Borrer (1781–1862) who noted the relatively loose and dishevelled flower-heads, with some of the flower-stalks longer than others and bearing a secondary head. It is found in more than twenty stations, most of them in Cornwall, though the individual colonies are small: there were only two plants near the church of St Enodoc. However, nearly fifty years ago it was first recorded in the Isles of Scilly, where it is now to be found in several different kinds of habitat on all the larger islands. (The late David Hunt who lived on St Mary's described it, somewhat disparagingly as 'looking like a cultivated leek that had gone to seed'). It has also been reported from Dorset and the remoter parts of Ireland.

However, Charles Babington (1808–1895), after whom the plant takes its name, came across it first in Guernsey. He was one of the many famous botanists associated with Cambridge University, but he was also an archaeologist, and also one of the founders of the Entomological Society (1833). In 1843 he published his *Manual of Botany* based on many expeditions up and down the British Isles. He was also founder, and for fifty-five years secretary, of the Club established to honour the name of another great Cambridge botanist, John Ray. Like Ray, Babington wrote a *Flora of Cambridgeshire* which was published in 1860, and a treatise on Brambles under the more formal title of *The British Rubi*. A year later he became Professor of Botany at Cambridge. Apart from the Leek, Babington – a stickler for creating new species from slender differentials – was commemorated in 'Babington's Orache' which is distinguished from its near relative the Hastate Orache only in the structure of the bracteoles. It is now included as a form of another orache with the botanical name of *Atriplex glabriuscula*. Such is progress.

IVh. *Oenothera stricta* Fragrant Evening Primrose

* 6th June Somerset

The Fragrant Evening Primrose is one especially attractive dune plant at risk – a naturalised immigrant. The first Evening Primrose greatly puzzled the wise men when it first travelled to England from its home in

North America. The eminent gardener and court apothecary, John Parkinson (1567–1650) was uncertain whereabouts to place the Tree Primrose which he called *Lysimachia lutea siliquosa Virginiana* in his famous work *Paradisi in sole Paradisus Terrestris* (1629).

'Unto what tribe or kindred I might referre this plant, I have long been in suspence' he wrote, 'in regard I make no mention of any other *Lysimachia* in this work: lest therefore it should lose all place, let me rank it here next to Dames Violets, although I confess it has little affinity with them.'

He is on firm ground there, in as much as the Dame's Violet, *Hesperis matronalis*, is a mauve-flowered member of the Cruciferae or Cabbage family.

Even though Parkinson grew the plant himself, it continued to puzzle him.

> Of the first yeare of the sowing the seede, it abideth without any stalke or flowers, lying upon the ground, with divers long and narrow pale green leaves, spread oftentimes round almost like a Rose, the largest leaves being outermost, and very small in the middle: about May the next yeare the stalke riseth which will be in Summer of the height of a man, and of a strong bigge size almost to a man's thumbe, round from the bottom to the middle, where it groweth crested up to the toppe, into as many parts as there are branches of flowers, every one having a small leaf at the foote thereof.
>
> The flowers stand in order one above another, round about the tops of the stalks, every one upon a short foot-stalke, consisting of foure pale yellow leaves, smelling somewhat like unto a Primrose, as the colour is also (which hath caused the name) and standing in a green huske, which parteth itself at the toppe into four parts or leaves, and turne themselves downewards, lying close to the stalke; the flower has some chives in the middle, which being past, there come in their places long and cornered pods, sharpe, pointed at the upper end, and round belowe, opening at the toppe when it is ripe into five parts, where in are contained small brownish seeds; the root is somewhat great at the head, and wooddy, and branched forth diversely, which perisheth after it has born seed.

Since then, the Evening Primrose has been assigned to the *Oenothera* family which includes a number of other choice garden plants such as the Fuchsias, Clarkias, Godetias, and the more humble Willowherbs. When literally translated *Lysimachia*, the botanical name chosen by Parkinson, commemorates Lysimachos, one of Alexander the Great's generals, who afterwards became king of Thrace. Literally translated, the word means Loosestrife, which even today is associated with a member of the Primrose family, Yellow Loosestrife – *Lysimachia vulgaris* – though not with Purple Loosestrife which adorns our streams with rods of purple and belongs to yet another family, the Lythraceae.

Today the Evening Primrose most often seen in the wild is the Large Evening Primrose, *Oenothera erythrosepala*, distinguished, for those carrying a hand-lens, by the red bulbous bases to its hairs and by its red striped sepals. Like Parkinson's Tree Primrose, this is a biennial but is not associated with Virginia and there is a strong belief that it may have arisen in Europe, even though, clearly, its presence there was unknown to Parkinson.

Oenothera stricta, the Fragrant Evening Primrose, which came here from beyond the Andes – either from Patagonia or Chile – is less common, and its numbers appear to be decreasing. It is more slender and more lissom than the commoner species so that the flowers appear larger in proportion to the plant. Though the botanical term *stricta* means 'strait' or 'narrow' i.e. 'upright', the plant seems no more or no less upright than its cousins of the same genus, though, perhaps it has fewer branches on the stem, and often none.

The leaves at the base of the plant are somewhat narrower and the flowers, yellow at first, turn reddish later, without losing their charm. There are probably a dozen localities where this beautiful species is still to be found. Those known to me were in Somerset, growing above the road on a strip of dune lying between it and the sea, but it would be worth looking for in any similar habitats in the south and west of the country.

IVi. *Lactuca saligna* Least Lettuce

One severely threatened species of the shingle is easily overlooked, namely *Lactuca saligna*, the Least Lettuce. This is not only because it is so much less in height (seldom more than 70 cm [28 ins] in this country) and very much less robust than the other lettuces, but also because characteristically the petals are closed and often appear in books as in real life as gold tips above the involucres. This is because they open only in bright sunshine. So if you are thinking of going to see a plant of Least

Lettuce on the other side of the county, ring the nearest market garden – or better still a kind friend – and get an assurance that the sun is going to stay out all day. The flower when fully open is hardly magnificent; in fact sometimes it shows a mere five or six pale lemon-coloured ray florets which would be put to shame by the nearest dandelion.

The striking features of the plant are its greyish-glaucous colour typical of so many other seaside plants, its white-ribbed stem – leaves undivided, arrow-tipped at the base and pointing to the skies – and the light green involucres beneath the flowers. There are no prickles.

Once it was known inland in Cambridgeshire, Huntingdonshire and Middlesex, and along the coast from Sussex and Kent to Essex, Suffolk and Norfolk. Today there may be fewer than nine sites shared between Sussex, Kent and Essex and, since 1982, it has been included on the Scheduled List of Threatened Plants.

John Ray discovered this plant in 1660 and named it *Lactuca sylvestris laciniata minima nondum descripta*, literally The Least Woodland Cut-leaved Lettuce not hitherto described. (The approved modern Latin-style to indicate a new but, as yet unconfirmed species, is now *sp. nov. mihi*). One of the two inaccuracies in Ray's description – the reference to cut leaves was amended in the third edition of Ray's *Synopsis methodica Stirpium Britannicarum* which read in part as follows: 'Lactuca sylvestris minima, the Least Wild Lettuce or Dwarf-Gum Succory. On a Bank and in a Ditch by the side of a small Lane or Grove, leading from London Road to Cambridge River, just at a Water-brook crossing the Road about a Quarter of a mile from the Spittle-house End.' (A.H. Exen and C.F. Prime the translators and editors of Ray's *Catalogue of Plants Growing Around Cambridge* suggest that this locality would be near the southern end of the present Coe Fen.)

The same entry added 'Tho. Willisel [Ray's friend] observed and shew'd it about *Pancras Church near London*. (By the path on the left hand side in the Close next this side *Pancras*, and by the Road Side in the same Close: Mr. Newton). By *Mr. Dale* it was found in the Eriffe [Eridge] Marshes in Kent.'

In its time the Least Lettuce has been found in many other places and, in the middle of the nineteenth century, it was present in nine vice-counties. In Cambridgeshire, Babington knew it and it was recorded at Earith, a village on the River Ouse, and as recently as 1951 on the Bedford River in Huntingdon, as well as in Norfolk and Suffolk.

The British Red Data Book of 1983 states that, since 1960, it has been observed in only nine sites in Essex, Kent and Sussex. Today the number may already be fewer. One classic site in Kent, where the Least Lettuce was to be found growing on the old sea wall at Cliffe, near Rochester, was drastically altered when the height of the wall was raised in connection with the construction of the Thames Barrage. For at least two years the Lettuce has not been seen there, and it seems also to have disappeared from its other Kent site at Seasalter in the Swale estuary.

In Sussex, a few plants are to be seen on shingle near Rye, but they have to be surrounded with close-meshed wire-netting to protect them from rabbits. This leaves Essex with the great majority of plants. There, the site is on the edge of a pasture sufficiently far from the Thames to be secure from any reconstruction of the sea-wall and the ridge has been declared an SSSI. The presence of cows there is considered on the whole to be beneficial. The reasons for the more general continuous decline in the population of the Least Lettuce is believed to be due to the climate, for Britain is probably the least warm and sunny of the countries in which it is found. Plants in the U.K. are confined to counties which are warm and southerly, and in them the Least Lettuce flourishes best on sites which are not only sheltered from the north but are inclined so as to obtain maximum benefit from the sun's rays. Where the Lettuce is protected from the north by a wall, it also derives benefit from the heat reflected from the stones behind.

Being an annual, it depends from year to year on the production of seed which is likely to be reduced in dull damp summers, raising the possibility of a continual reduction in numbers. Dispersal is also hindered by rain which prevents the 'parachute' carrying the seeds from opening properly. Following a poor year, the number of places in which new seedlings could establish themselves will have been further restricted in the meantime since perennial plants will have taken over the available bare patches of ground. Grassy fields along the saltings are improved (for the Least Lettuce) during the winter if some of the perennials are trodden or grazed out of existence. Even so, the survival of the Least Lettuce may depend on the fact that it can tolerate a more saline environment than some of its competitors.

IVj. *Petrorhagia nanteuilii* The Childling Pink

One rare member of the Carnation family makes up for its modest style with a tongue-twisting name – *Petrorhagia nanteuilii* – the Childling

IVa.

△IVc. ▽IVd.

△IVb. ▽IVe.

IVa. *Matthiola incana* (ht 30-60 cm)
IVb. *Aster linosyris* (ht 40 cm)
IVc. *Ophrys sphegodes* (ht 20 cm)
IVd. *Orobanche caryophyllacea* (ht 25 cm)
IVe. *Trifolium stellatum* (ht 15 cm)

IVf.

IVg.

△IVk. ▽IVm.

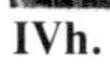

IVh.

IVf. *Matthiola sinuata* (ht 60 cm)
IVg. *Allium babingtonii* (ht 180 cm)
IVh. *Oenothera stricta* (ht 90 cm)
IVk. *Lathyrus japonicus* (ht 30-40 cm)
IVm. *Inula crithmoides* (ht 50 cm)

Pink. It is usually less than 30 cm (12 ins) high with a grass-like unbranched greyish-green stem topped by a large pale brown papery bract from within which the flowers, pale pink, with notched petals emerge, most often one at a time – although a bouquet of three at once is sometimes achieved. They tend to close in dull weather, making the plant well-nigh invisible towards sunset.

This is a high-risk species since it has disappeared from the two stations in Hampshire where it grew up until 1965 and, apart from the Channel Islands, is now to be found only in West Sussex (according to the *Sussex Plant Atlas* – at Shoreham and around Pagham), in Bedfordshire and at two stations in south Wales. One of these is on the embankment of a disused railway line to which, it is suspected, seeds were carried with imported iron-ore.

Being an annual it appears in unpredictable spots on different sites, but seems at home equally on fixed dunes, sandy scrubland and open shingle on which in Sussex 1977 there was a stand of more than thirty plants, closely packed, completely isolated from pedestrians and rabbits alike, and without competition from any other vegetation. But in a vast expanse of shingle it was hard to mark the spot and the next year the pinks were not to be seen. As an annual it is dependent on good seed production every year and is more vulnerable to gormless pickers than if it were a perennial. However, since the main flowering period is in mid-June, the plant escapes some of the holiday pressure.

Earlier writers lumped the Sussex flowers together with another rather similar species, *Kohlrauschia prolifera*, Proliferous Pink, and it was not until 1962 that the Sussex plants were recognised as *Petrorhagia nanteuilii*, recorded for the first time in Britain. It has since been concluded that other previous sightings of similar small pinks in the Isle of Wight, Kent, Middlesex, Berkshire, Buckinghamshire and Norfolk should, strictly speaking, be taken as referring to *Kohlrauschia prolifera*. It is probably significant that, whereas *K. prolifera* behaves like a casual in Britain, *Petrorhagia nanteuilii* has persisted.

Petrorhagia nanteuilii could well have been the plant that John Gerard described as *Armeria prolifera*, Childling Sweet Williams, with a convincing illustration and a note that it was called by some, *Caryophyllus sylvestris*. John Parkinson, absorbed with his garden plants, gives no space to the wild species but the third edition of Ray's *Synopsis Methodica Stirpium Britannicarum* (1724) states that *Carophyllus sylvestris prolifera* was found by the Reverend Mr (Thomas) Manningham (1684–1750) on 'Selsey Island' (lying to the west of Pagham and since connected to the 'mainland' by a causeway), and until quite recently it has been reported still to be growing on shingle at the eastern end of the former 'island'.

Petrorhagia is derived from the Greek for 'Holy Rock', but I must plead ignorance as to whether Nanteuil refers to Robert Nanteuil, the seventeenth-century French engraver; to Celestin Nanteuil the nineteenth-century artist and lithographer, or to either of two Nanteuils in Northern France. Neither of them is credited in guide books with any such attraction.

IVk. *Lathyrus japonicus* Subspecies *maritimus* Sea Pea

* 8th July East Sussex

There is something rather voluptuous about the way this plant strews itself across the beach. Its leaves are softly rounded – glaucous green – almost paddle-shaped and matched close together like beads on a necklace. They offer a perfect background for the large, brilliant blooms which vary, according to the light, between blue, mauve and deep rose, with an intensity often lacking in the garden Sweet Pea.

This plant has succeeded in establishing itself in Arctic territory all round the world – from north of Leningrad through to Siberia and China and across to North America.

In Britain, however, it is very local, occurring in a few counties along the south coast, in Aberthaw (Glamorgan) and Bardsey island (Caernarvonshire), in East Anglia and in Angus, and West Ross where, however, the plants are a narrow-leaved form. On the south coast, those wishing to see this beautiful plant should search the shingle to the east of Winchelsea – although here as elsewhere the beaches are under permanent threat of development.

There are also colonies on the beach of the nature reserve at Rye Harbour in East Sussex where, however, the warden should be consulted before entry. The allusion to Japan in the botanical name refers, probably, to the fact that the plant is found in the southern part of Sakhalin Island which formerly formed part of Japan, was occupied by Soviet forces during the Second World War, and is still a matter of dispute between the two countries.

IVl. *Mertensia maritima* Oyster Plant

The Oyster Plant belongs to the same family as that popular border perennial Soldiers and Sailors, alias Pulmonaria or Lungwort, and is almost a seaside form of it. The flowers, like those of Lungwort, are pink at first, but later turn partly blue. The leaves of the Oyster Plant, however, are glaucous, unspotted and, like so many other seaside plants, fleshy.

Most of the Oyster Plant rests on the shingle, with only the new growth rising up a little. This is, no doubt, due to the exposed conditions in which it grows but it is unfortunate, for the plant has a somewhat shabby and neglected appearance.

This is on the whole a north-country plant, the best-known site being in the area which was once known as Wigtownshire, but which is now Galloway and Dumfries. However it has not escaped notice that the Queen Mother, when in residence at Mey Castle near John O'Groats, personally took some honoured guests down to the shore beneath the castle to see the Oyster Plants growing there.

The plant derives its name from a widely held – but seldom tested – belief that the leaves when eaten possess a flavour akin to that of oysters.

IVm. *Inula crithmoides* Golden Samphire

* 8th September West Sussex

This is a rather beautiful plant with golden-orange daisies about an inch across and dark green fleshy stalkless leaves. It is never far from the sea waters and indeed seems to prefer a spot washed by the tide, so that the name Samphire – derived from Saint Pierre who walked on the waters – is not inappropriate. Unfortunately the leaves are beloved by rabbits which means that where there are dunes, the plant is at risk. On salt marshes, cliffs and rocks, prospects are better and large colonies are sometimes found, but grazing and pollution remain as permanent threats.

At one well-known site in the south – East Head on Chichester Harbour – Golden Samphire has disappeared completely during the last ten years, although it persists at another less accessible salting further north.

This is a late-flowering plant rarely on show before mid-July; it continues blooming into the early autumn. In 1975, the year of the drought, when many plants finished flowering a month before their time, *Inula crithmoides* was still in flower in the second week of September. It has even been seen as late as October. The botanical name *crithmoides* suggests the resemblance between Golden Samphire and Rock Samphire, since both have fleshy edible leaves and both cling to the seashore. Botanically speaking, however, Golden Samphire is no relation to *Crithmum maritimum*, Rock Samphire, which belongs to the Carrot and Cow-Parsley family.

— V —

Plants in Fear of Drying Out

* (See following p. 146 for illustrations)

Va. *Bartsia alpina* Alpine Bartsia

* 30th June County Durham

A sombre plant this, of moist grassland and damp rock ledges, with flowers of dusky purple and purplish bracts beneath to match. The flowers are two-lipped with the upper one projecting forwards like a hood – or perhaps a monk's cowl, an impression heightened by its velvety texture. From within the cowl there appear white stamens – the ends of two miniature white wax tapers. The leaves, bristly and dark green, are scalloped with saw-like teeth and are said to be decussate, that is they grow in pairs on opposite sides of the stem, but each pair is at right angles to the next.

Alpine Bartsia was first described in Britain in 1668 by John Ray who found it growing near Orton, then in Westmorland but now in Cumbria. It also flourishes in another non-alpine habitat in Upper Teesdale, Co. Durham, on a slope so close to the Pennine Way that many a knapsacked hiker has passed on his way a few yards above it, unconscious of the botanists just out of his sight. Slopes with well-watered soil are the preferred habitat for this plant, which also occurs in Yorkshire, Perthshire and Argyll. In all there are probably between 20 and 30 colonies.

The genus Bartsia commemorates the name of the German botanist Johannes Bartsch (1709–1738). It came about in this way. Linnaeus, in 1737, had been offered the post of Botanical Director to the Netherlands Chartered Trading Company in Surinam, but rejected the offer feeling that he would not be suited to the tropical climate. Instead he recommended his friend Bartsch for the task. Within six months, however, Bartsch died of a tropical fever. Linnaeus blamed himself, and sought to make amends by immortalising Bartsch's name in a plant dressed in mourning colours.

Vb. *Damasonium alisma* Starfruit

* 7th July Surrey

This plant could well have been extinct in Britain before these words had appeared in print, for the number of starfruit plants has decreased alarmingly quickly in recent years. It was saved, at almost the very last minute by a rehabilitation of the only pond on which it was known to grow. An earth-mover was needed to remove the mud and clean the willows and the 'new' plants are well below average size. But at least they have flowered and fruited. Starfruit was already 'very local' in 1962 according to the second edition of Tutin, Clapham and Warburg's *Flora of the British Isles* which, however, listed it as occurring in Hants, Sussex, E. Kent, Surrey, Berks, Bucks, Middlesex, Hertford, S. Essex, Worcester, Shropshire, Leicester and South-East Yorkshire. But only eight years later it was reliably reported from only one site – a small pond in Surrey. Some experts believe that the decline in its numbers has been due to pollution, drainage or to the deliberate infilling of ponds. Others are certain that failure to keep a clean, mud-free gravelly floor to the pond is the real reason, and that paying heed to this requirement accounts for its revival.

The flowers of *Damasonium alisma*, like several other water plants (e.g. Frogbit, Water Plantain, Water Soldier and Arrowhead) consist of three whitish petals arranged like the blades of a propeller. Each bears an orange spot. What distinguishes it, however, from other similar plants is the 'fruit' consisting of six carpels radiating from a common centre like the rays of a star.

Vc. *Epipactis palustris* Marsh Helleborine

* 14th July North Devon

Public pressure, drainage of marshes, and instability of sand dunes are the three main threats to this beautiful orchid – although the degree of risk is nothing like so serious as with the previous plant, Starfruit. Indeed there are colonies of Marsh Helleborine such as those in Braunton Burrows, near Barnstable, Devon, where the plants spread so widely in the damper slacks through rooting stems, and grow so thickly that it is sometimes difficult to avoid treading on one.

The typical bloom consists of five sepals and petals coloured light brown or purplish outside and pale reddish purple within. Below them is a white lip marked with yellow and ruffled like a lace cravat.

There is also a yellowish-green form, *Epipactis palustris var. ochroleuca* with a white lip.

This species seems to prefer the wide-open spaces, and dunes composed of calcareous sand derived from sea-shells. The flowering period is a lengthy one, and one flower can usually be seen as early as June and as late as September.

Vd. *Gentiana pneumonanthe* Marsh Gentian

* 25th August East Sussex

This beautiful plant with its 5 cm (2 ins) trumpets of deep blue pointing skywards has recently become very local – particularly in the south. It was formerly known in Kent, but not now. Half a century ago there were thirty records in Sussex but more and more of them have become non-recurring. More recently the number of sites was put at ten, but at least one of these to my knowledge has vanished – probably due to overgrowth by gorse. Equally damaging, however, have been the efforts to control the gorse by burning, which incidentally encourages the growth of bracken.

Ideally, this gentian prefers an acid soil and a damp heath. But often trampling – including that by the beloved New Forest ponies – compacts the ground so that it is no longer capable of holding the water that the plant needs. Some grazing on the other hand can be beneficial. Other hazards to heaths beloved of the Gentian include motor-cycle scrambles, picnic fires, ploughing, invasion by trees, conversion to recreation grounds and other forms of development. The counties of Suffolk and Dorset have suffered particularly.

Those seeking to photograph the flowers of the Marsh Gentian should avoid taking front or side-lit shots in which the light is reflected from the petals back to the camera. Shots taken *contre-jour*, that is with the camera pointed towards the main sources of light stand a better chance as the light will then pass through two layers of blue petal before reaching the camera. Ideally the picture should show, too, the green lines ornamenting the outside of the corolla.

Ve. *Hammarbya paludosa* The Bog Orchid

* 17th July Hampshire

Most botanists will admit that the Bog Orchid – *Hammarbya paludosa* – is exceedingly hard to find – even when one knows where to look for it. In the south, the New Forest is one good patch. One reason – it is to be hoped – is that God-fearing botanists hesitate to impair the kind of bog it grows on by stepping too close to it. One wishes that it might be possible to warn off those delightful New Forest ponies too. One could argue that they do far more damage to some plants (though not to all) with their hooves than the botanist with his Wellington boots.

It is, of course, our smallest native orchid with a stem of 12.5 cm (5 ins) or less, and perhaps we should all look for it wearing special glasses focussing at a distance of say 15 metres. Then again both the stem and the flowers are greeny-yellow, but the flowers are just a shade lighter than the stem, breaking the plant effectively into two parts according to the approved camouflage principle. Besides which, seen at a distance, the flowers, lying along the stem, seen together suggest to the uninitiated one of the sedges, and so often this turns out to be the case that one feels like giving up looking for the orchid.

There are of course times when one's eye is in and days when it is not. How often has one visited a place where one has been told to look for such and such a plant, and has come away without seeing it? Too often, probably. But on good days or bad days alike, it pays to start out with a very clear picture of the plant you hope to discover. Undoubtedly botanists who have been to a site before have an advantage; for it is a well-known maxim that 'you find what you expect to see'. The eye looks for the image to which it is accustomed – just as when reading a text, one unconsciously overlooks a misprint and accepts instead the more familiar word that was intended. And from a list of random names one's own seems easier to recognise than any other.

And now, having discussed all the excuses for not finding it, let us examine the plant itself. The individual flowers have all the curves and shine of sphagnum moss which they match so perfectly. But they have one peculiarity. In most other orchids the lip or labellum which ends up at the bottom of the flower originally developed at the top, and the whole flower has turned through 180 degrees so that when it is fully developed its north became its south. In the Bog Orchid, however, the flower has made a complete turn, so that the lip which started at north turned through south and eventually completed a full turn, ending up at north again.

The lip seems small in comparison with the two sepals which flank it like a pair of rabbit's ears. The third sepal points downwards and the remaining two strap-like petals, one on each side of the flower, are relatively insignificant.

The absence of a labellum on which insects can land is, however, no disadvantage for the Bog Orchid as it reproduces itself by means of small bulbs which form at the top of the leaves.

One vexing characteristic of the Bog Orchid is that it frequently fails to appear in the places where it appeared twelve months ago. But in the habitat it favours one place looks so like another that it is difficult to be perfectly sure. But it is nevertheless quite as scarce on the continent of Europe as it is in Britain – perhaps because they drain marshes there too.

Vf. *Liparis loeselii* Fen Orchid

* 11th July Mid-Glamorgan

In the days before this orchid was known to grow in Wales, the description 'Fen Orchid' seemed highly suitable. More recently, however, it has been discovered in six 'burrows' in Glamorgan and three 'burrows' in Carmarthen. ('Burrows', but the way, is a recognised term for an area of mixed sand dunes and slacks, the latter being small shallow depressions). It is in the slacks on the burrows that the western 'Fen Orchids' grow – in the site that I know best there are two colonies, each surrounded by acres of solid ground and creeping willow.

Fen Orchids in general have but two leaves, each pale green with an edged keel beneath, shiny as plastic. They grow almost, but not quite, opposite one another, giving the plant a lopsided look. The leaves of the Welsh plant are, however, blunter and more elliptical than those of their eastern cousin and the Welsh plants are often alluded to as *var. ovata*.

The flowers – pale straw-coloured rather than, as sometimes described, greenish – are somewhat irregular with the petals and sepals in wild confusion and with only the broad central lip hanging vertically downwards.

In the Fen Country the main threats to this orchid come from changes in the water-level due to mechanically dug drainage, enrichment of the water through run-off of farm fertilisers containing nitrates and phosphates and, where permitted, disturbance by anglers and motorcraft. More recently special measures have been taken to preserve minimum water levels in the area in some cases by means of sluices and in others by lowering the soil level.

Vg. *Ludwigia palustris* Hampshire Purslane

Ludwigia palustris, now known as Hampshire Purslane, is hardly in danger of being picked as an adornment for the drawing-room flower vase, and would happily slip away from view underwater if allowed to do so. Its roots lie in shallow pools, and the plant flourishes both below and above water. The stems, round in cross-section, are weak, hairless and reddish, and those leaves beneath the surface of the water are reddish too, though watercress green where they poke up into the open air.

For a plant that belongs to the same family as the Evening Primrose and the Fuchsia, *Ludwigia* is astonishingly unpretentious. The flowers, minute, almost stemless, are tucked into the axillae of the leaves. They are often without petals, in which case all that can be seen are the tips of the four pale green sepals spreading outwards.

Ludwigia is easily confused with a much more common species which is often found growing alongside it but which belongs to a different family: namely *Peplis portula*. This commoner is nevertheless known as 'Water Purslane'. It, however, has angled stems (not rounded as in *Ludwigia*) and six sepals instead of four. The leaves, usually in pairs opposite each other, are shaped like the blade of a paddle, whereas the leaves of *Ludwigia* are sharply pointed at the tip. The *Ludwigia* leaves are also arranged in pairs, but the pairs are described by botanists as being decussate, that is, each pair stands at right angles to the pairs below or above it.

At one time, the men of Sussex would have strongly contested the claim of Hampshire to be the proprietors of the 'English' name. During the first thirty-seven years of this century, Sussex botanists relied on a slim *Sussex Flora* prepared by the Rev. F.H. Arnold who lived at Hermitage, Westbourne, in the western extremity of a county stretching almost eighty miles to the east. The first edition of this work was published in March 1887 at a time when Surrey and Hampshire already had their own *Floras*, and after Arnold himself had studied the flowers of Sussex for a quarter of a century. A second edition was prepared and issued in 1907 with a preface dictated by Arnold to his daughter Marian on the morning that he suffered a stroke from which he died a few days later. Arnold took a firm line on what plants he wished to accept as Sussex Flora. 'I would first here protest against the growing practice of including in our lists many foreign species, and thus causing confusion. Not only garden escapes are admitted, but the sweepings of corn and oil mills, of stores containing hay seeds, also where foreign grain is landed, and even rubbish heaps are ransacked, with the view of coming upon aliens, some of which are by no means desirable ones. Such species are usually non-permanent, our climate being uncongenial, and I exclude them altogether.'

At this time, Ludwigia still grew in Sussex both in the eastern part of the county at Buxted 'in an old iron pit', where it had been found by William Borrer (1781–1862) in 1827. Borrer sent specimens of the plant to a fellow botanist named Collins in 1847 who sent them on to Arnold. But Arnold added that the plant was probably near extinct.

Wolley-Dod's *Flora of Sussex* was the next survey to be published, and ran to more than 500 pages compared to Arnold's 154. Wolley-Dod admitted that the plant was 'probably extinct'. The spot in which it was reported to grow 'between Buxted bridge and the Tanyard' appeared to have become overgrown, he wrote, and another siting reported from the Little Ease Millpond at Cuckfield was to be regarded as unreliable.

The more ambitious *Sussex Plant Atlas*, compiled by P.C. Hall from records collected between 1966 and 1978 by the Sussex Flora Society, was published in 1980 by the Borough of Brighton's Booth Museum of Natural History. It adds the laconic note 'Extinct. Last seen ca. 1876.'

Ludwigia is still to be found in several shallow ponds in the New Forest, and has also been recorded in Epping Forest. Provided that the sites do not become overgrown or are filled in in the name of progress, it should continue to flourish on its present stations. Indeed in some of these it is the dominant species.

Two of the sites on which it occurs have been declared as SSSIs.

Vh. *Pulicaria vulgaris* The Lesser Fleabane

* 16th Sept Hampshire

This humble annual, with its greyish woolly leaves and daisy-like flowers of tarnished gold, is a relic of the Middle Ages from the days when every village had its pond, the borders of which were frequented and fertilised by the domestic duck. In the winter, when the ponds were at their fullest, and the waters reached their extreme limits, any perennial plants which might interfere with spring seedlings of *Pulicaria* were submerged and their leaves dissolved. Then, in early spring, the ducks would continue to puddle in the mud keeping it more or less bare as the waters retreated to their summer level, and the seeds of

Pulicaria vulgaris, dispersed the previous autumn, could germinate more or less unhindered close to the water's edge.

The French botanist Mathias de l'Obel first recorded this plant in England in 1570 – the same year he published his most famous book *Stirpium Adversaria Nova* (A New Notebook of Plants).

Barnard's Green, where de l'Obel made his discovery, has now become a part of Greater Malvern, a transformation of the kind which must have happened in many other counties, and one which helps to account for the present rarity of this species. In many a village the pond has disappeared altogether, having become overgrown, silted up or intentionally filled in, often with unsuitable rubbish. Even earlier the ducks had vanished. Motorists and visitors with their dogs unleashed had seen to that. Even decaying leaves will, in time, render a pond too acid for normal plant life.

By 1833 *Pulicaria vulgaris* had vanished from Cambridge where it had long been known. In 1866 it was described by Boswell Syme in Sowerby's *English Botany* of that year as 'rather rare'; today there are fewer than a dozen sites, though the inconspicuous character of the plant and the large number of available mud puddles in the countryside at large make it possible that at least some more plants will come to light.

The largest of the present sites, and the area with the greatest potential is the New Forest, where the wild ponies perform the same function as the ducks used to in earlier centuries, by poaching (i.e. pock-marking) the mud around shallow depressions which hold water during the winter, and by spreading the seeds from one location to another.

The modest stature of *Pulicaria vulgaris* – up to 45 cm (18 ins) – and its much branched stem help to distinguish this plant from its more flamboyant relative *Pulicaria dysenterica*, Common Fleabane, to be found in almost every other ditch.

The flowers of the rare plant have but one row of rays round their edges and these rays are hardly longer than the tube-like flowers in the central disc.

Vi. *Teucrium scordium* Water Germander

Rivulets, ditches and small ponds are the favourite habitat of this humble plant and the gradual disappearance of these refuges accounts, probably, for the fact that Water Germander which was formerly known in Berkshire, Oxfordshire, Suffolk, Norfolk, Huntingdonshire, Northamptonshire and Yorkshire, is now found in two counties only – Devon and Cambridgeshire.

F.H. Perring, P.D. Sell and S.M. Walters in their *Flora of Cambridgeshire* noted that it was very intolerant of competition from taller marsh-growing plants, and therefore never persisted for long in any one site unless the ground had been cleared for clay or peat digging.

This theory is borne out in a locality in north Devon where it flourished on a site which was cleared during the Second World War by the U.S. Forces, then practising for the invasion of Normandy, who wished to make a road there. Even so it needed the warden of the site to part the vegetation before one could see Water Germander growing beneath. The plants were perhaps 15 cm (6 ins) tall, with coarsely cut leaves and the typical Germander flower consisting of one lower-lip only coloured pinkish-purple.

Vj. *Saxifraga hirculus* Yellow Marsh Saxifrage

It is all too easy to confuse the two yellow saxifrages which ornament our mountains and moors. The Yellow Mountain Saxifrage, *Saxifraga aizoides*, is much the commoner. It sends up many sterile shoots, has stalkless leaves and flowers with petals too narrow to conceal the green sepals behind them. Most plants have flowers in loose clusters, the older flowers in the centre, the younger ones round the edges.

The Yellow Marsh Saxifrage, on the other hand, has leaves stalked and partly sheathed, reddish hairs on the stem, rounded petals a centimetre or more long, twice the size of *S. aizoides* and broad enough to conceal the sepals. Usually, though not invariably, the flowers are solitary. The Yellow Mountain Saxifrage favours a stony niche close to a stream often not far from the roadside. But the Yellow Marsh Saxifrage keeps to the hollows between clumps of rushes on damp moorland hillsides – usually in undisturbed areas, often so remote that it is difficult to relate them to any landmark or map reference. The numbers of plants in flower varies from year to year. Do not expect any blooms before mid-August.

The main threat to the plant comes in this case from misconceived and ineffective attempts to drain the moors by scarring them with deep mechanically cut ditches.

Vk. *Mentha pulegium* Pennyroyal

Anyone fortunate enough to find this Mint can confirm his good luck by noting the characteristics which distinguish it from other species of the same tribe. First of all it is prostrate in habit or nearly so; the leaves, too, are smaller than those of other mints (except for those of the minute mat-forming *Mentha requienii*, known as Corsican Mint) and they are scarcely, if at all, indented. The neat pin-cushions of mauve flowers are widely distanced from one another, and the main stem joining them is seen to be relatively stout.

The rapid disappearance of this plant from the 55 vice-counties where it was formerly known is probably linked with the decline of the village pond, no longer used as the villagers' water supply (even cattle now expect piped water), and so often filled in or allowed to become overgrown. The total number of sites for Pennyroyal may now be as low as fifteen. Unstability of water level, failure to clear the pond, and drain-off of nitrates and phosphates no doubt will kill off some plants each year.

According to the Rev. C.A. Johns in that classic work *Flowers of the Field*, Pennyroyal was extensively cultivated in his day in cottage gardens to provide mint tea, then a trusted remedy for colds. To the Romans, however, Pennyroyal was *puleium* or *pulegium*, the plant that was the bane of *pulex*, the flea. The name became *puliol* in Ancient French and the adjective *real* was added after the plant had found favour with royalty. But no satisfactory explanation is offered as to just how puliolreal was transformed into Pennyroyal.

Vl. *Fritillaria meleagris* Fritillary

* 2nd May Gloucestershire

A stand of Fritillaries growing in the water-meadows near the source of the Thames must be one of the most idyllic sights that a botanist could wish for. It is hard to resist walking up to admire each and every one of the flowers, varying in colour from the normal deep purple through pale pink to blooms of a skimmed milk complexion with the petals marked outside near the base with a light dash of green. The darker petals are bizarrely marked with a chequered pattern which resembles the one used by the Romans on their dice-boxes for which their word was *fritillarus*. Before the flowers open, the regular chequered pattern is distorted so that you see an array of purple lines scrawled across each pallid bud as though it were wire-netting squashed sideways. The buds are long, flat and narrow, and pointed at the end and the markings evidently suggested to someone those of a reptile's skin – hence the alternative vernacular name for the plant – Snake's-head.

The stems on which the blooms are carried rise vertically and then arch over at the top like water from a fountain so that the flowers hang downwards like the bells in an old-fashioned manor house.

One would naturally suppose that such an unusual flower would have attracted the early attention of botanists, yet it was not reported until 1736, and this has led to some doubts as to whether it is really a native plant or one that has been introduced to the wild via the wheelbarrow.

On the other hand, there is no positive evidence to show that it has been introduced to any of the sites where it has grown for centuries. These are scattered from Gloucestershire, through Oxfordshire and Berkshire, in the upper Loddon valley as far east as Suffolk where the Fox Fritillary Meadow near Framsden, under the protection of the Suffolk Trust for Nature Conservation, offers a wonderful carpet of fritillaries with an estimated population of nearly 50,000 plants to the acre. This meadow which is in a small valley through which runs a tributary of the river Deben, is not cut for hay until mid-July, that is after the Fritillaries have seeded.

Another excellent reserve at Framsden is on private land within a farm, and permission as in the case of the previous site can be obtained in advance through the Suffolk Trust. Mickfield Meadow near the village of that name is another SSSI site under the protection of the Suffolk Trust.

Nearer London it may be possible to get permission from the Duke of Wellington's Agent to visit the fritillary meadow on the Duke's estate at Strathfield Saye Park, near Wokingham, thus combining a stately home with a stately flower. The Fritillary grows only in meadows that have not been ploughed, improved with fertiliser or drained. Most of the meadows that have survived have peculiarities which make them unsuited to intensive agriculture, either because they are flooded in winter, hard to drain or too rough or steep for tractors. They are best preserved when grazing is permitted only before May or after the hay has been cut in late July or August.

— VI —

Some Hangers-on – Some Casuals

* (See following p. 146 for illustrations)

VIa. *Aristolochia clematitis* Birthwort

* 8th June East Sussex

It will be appreciated that the *British Red Data Book* deals mainly with those which are *native* to Britain or probably so. But some handsome plants have been introduced from abroad and are at risk precisely because they are aliens and therefore not at home.

Aristolochia clematitis, Birthwort, is one of these, for its natural home is in central and southern Europe. Surprisingly – for it has an unattractive smell – it has long been treasured by herbalists. According to John Gerard 'Dioscorides writeth, that a dram weight of long Birthwoort drunke with wine and also applied, is good against serpents and deadly things: and that being drunke with myrrhe and pepper, it expelleth whatever is left in the matrix after the child is delivered, the floures also and dead children and that being put up into a pessarie it performeth the same.'

Long Birthwort referred to by Dioscorides and described by Gerard is burdened with 'leaves of an overworne greene colour and a grievous or lothsome smell and savour' and the main difference between it and *Aristolochia clematitis* is that the latter climbs, whereas Long Birthwort sprawls. In both, the leaves are heart-shaped, and the tubular flowers are of a dullish pale yellow. A flap above the mouth of the tube suggests the flattened head of a cobra poised to strike.

Aristolochia clematitis is a rare plant, surviving most often near ancient monasteries or priories from whose infirmaries it must have been prescribed. As if to prevent self-pollination the ovary in each flower becomes ripe before the pollen for it has developed in the anther. The flowers are pollinated by small flies which are trapped in a chamber at the base of the flower until the stamens have ripened.

Gerard also describes *Aristolochia rotunda*, Round (-leaved) Birthwort, sometimes called Smearwort, sometimes Dwarf Birthwort, which

△Vb. ▽Vd.

△Va. ▽Vc.

Va. *Bartsia alpina* (ht 20 cm)
Vb. *Damasonium alisma* (ht 20 cm)
Vc. *Epipactis palustris* (ht 40 cm)
Vd. *Gentiana pneumonanthe* (ht 20 cm)

△Vf. ▽Vl.

△Ve. ▽Vh.

Ve. *Hammarbya paludosa* (ht 10 cm)
Vf. *Liparis loeselii* (ht 15 cm)
Vh. *Pulicaria vulgaris* (ht 20 cm)
Vl. *Fritillaria meleagris* (ht 50 cm)

△VIa. ▽VIc.

VIb.

VIe.

VIa. *Aristolochia clematitis* (ht 90 cm)
VIb. *Phyteuma spicatum* (ht 85 cm)
VIc. *Inula helenium* (ht 150 cm)
VIe. *Centaurea calcitrapa* (ht 60 cm)

△ **VIf.** ▽ **VIh.**

VIg.

VIf. *Centaurea solstitialis* (ht 60 cm)
VIg. *Campanula rapunculus* (ht 60 cm)
VIh. *Himantoglossum hircinum* (ht 45 cm)

has similar flowers but of a dull brownish purple hue. This plant, too, has been naturalised in Britain, but has survived so far as I know only high on the downs at the edge of a wood at South Hawke, in Surrey, where it has been known for more than sixty years. One wonders how it could possibly have been introduced there. Did it, perhaps, grow there naturally as a native, if very rare, plant?

Indeed, the whole subject of alien flowers – all risk-takers at the time they first appeared in Britain – is surely a fascinating one. Medicine is only one of the agencies that brought them to this country. Other interesting aliens have been imported accidentally in shoddy, the shredded woollen rags used for making up cloth, and also in bark from the Valonia Oak used in Britain in the tanning industry. The 'shoddy' aliens often turn out to be plants with fruit, equipped with hooked teeth such as members of the Xanthium (Cocklebur) genus. Canadian Fleabane, a relative of *Pulicaria vulgaris*, Lesser Fleabane, described earlier in this book, is believed to have reached Britain as part of the filling of a stuffed bird. Other aliens arrived in the ballast of ships.

One more alien, *Dianthus caryophyllus*, Clove Pink, which grows on the walls of Rochester Castle was said to have been introduced accidentally from France in the mortar used by the Normans who built the fortifications.

Some aliens including the gaudy *Impatiens glandulifera*, Himalayan Balsam (sometimes Policeman's Helmet) just stepped over the garden wall in the middle of the last century, or was it carried over to the nearest ditch in a wheelbarrow? Today it is at home on hundreds if not thousands of small waterways and far more at home than many of our natives.

This is indeed one way in which aliens are distinguishable from natives. They either prosper inordinately or they die out. Many survive for a generation only – and these are often described as 'Casuals'.

Those that prosper often establish themselves in waste places or open spaces for example, along the edges of streams, as in the case of Policeman's Helmet and the yellow Monkey Flowers, or on roadside verges.

VIb. *Phyteuma spicatum* Spiked Rampion

* 12th June East Sussex

This has a tall 76 cm (30 in) ramrod-straight stem, at the end of which is an upright cone of closely packed flowers resembling a vertical sprig of greenish white Buddleia. It is described as native to this country, and

Gerard in 1596 included the Great Rampion – as he called it – in a section on wild plants. However, we know that the root is swollen and fleshy, and that John Parkinson writing in the seventeenth century declared that its roots were used for 'sallets, being boyled and eaten with oyle and vinegar, a little salt and pepper'. So it was a garden plant even then. Later it is known to have escaped from gardens in Warwickshire, Staffordshire, Merioneth, Derbyshire and in Scotland.

In Sussex where it is supposed to be a true native, it was not mentioned until 1824, which seems strange, and one of the show sites in Sussex is in Abbot's Wood, downstream from Michelham Priory where it could easily have contributed to the monastic diet.

At large it is a native of central and southern Europe, but it has certainly made itself at home not only in copses but in verges here and there along a score of lanes and minor roads of East Sussex.

VIc. *Inula helenium* Elecampane

* 30th August East Sussex

A fine plant this, and capable of growing to nearly 150 cms (5 ft), thus generally dominating its surroundings, whether in fields on verges or on the edge of some copse. It sometimes appears alongside a hedge where, however, it will be under threat from spray drift, stubble-burning or even chemical waste dumped in the hedge-bottom.

Even the verges in the countryside are not safe territory, particularly where they are level with the highway. There, where the hedge has been cut down, the ditch beside it is filled in and part of the verge is ploughed up. Also where roads are often narrow and farm machinery wide, the verge forms a convenient lay-by for leviathans to crawl past one another. Farmers stand their spraying equipment on the verge before loading it up and again when cleaning down.

Where the ditch is left, large parts of the verge are cut through to allow water to run off the road into the ditch; in the old days a spade's width used to be considered enough for this purpose. Today earth-movers do the job.

Sometimes when the stubble is burnt, the verge is scorched as well. Apart from this, large stretches of verge are dug up in order to lay or maintain telephone and electricity cables, and gas and water mains, often without notice to those interested in conservation.

Elecampane is regarded as an introduced plant and not native. The herbalists of old found the roots useful for inducing perspiration to banish fevers, for clearing the throat of phlegm, and as a diuretic for patients suffering from liver or kidney complaints. The leaves were used to add flavour to the pot and the stem as a base for candy.

To botanists of old this plant was *Enula campana*, but when it was observed that it differed from other *Enulas* in yielding a 'fruit' with only four sides, it was renamed *Inula*. 'Elecampane' was a corruption of *Enula campana*; Inula, however, was the form used by Horace.

VId. *Buxus sempervirens* Box

Here is another of those plants which may be part native or part introduced.

The native bushes, it has been suggested, are those that grow only on bare chalk and limestone on sites remote from gardens where no one is likely to have planted them – a typical example being the box bushes on the steep slopes of Box Hill in Surrey. Garden escapes – or bird-sown bushes – would not, presumably be confined to chalk or limestone.

We know that the Romans used evergreen Box plants in their funeral rites and that they would have known the plant in Italy. But whether they introduced their own plants or used local native stock must remain open to question.

Box flowers are single-sexed, but male and female flowers grow in the same leaf axillae, and the stamens of the male flowers give a dash of yellow to the petal-less whitish green inflorescence.

The Box edging used in gardens is, of course, a dwarf variety.

VIe. *Centaurea calcitrapa* Red Star-Thistle

* 13th July East Sussex

Gilbert White was the first to mention this unusual plant and referred to it in his *Garden Kalendar* (1765) as 'a thistle with an echinated head and little down to ye seeds'. (Echinated, by the way, is – or was – a botanical term denoting 'furnished with bristles or prickles' and derived from the Greek word used impartially for 'hedgehog' or 'sea-urchin'). White's journey to Ringmer across the sort of country in which this thistle was – once upon a time – fairly common, provides fairly conclusive proof that

C. calcitrapa was the plant he meant. Gilbert White's statement that the seed had little down on it was an understatement. The achene is without hair.

The spines around the flower of this plant are an extension of the bracts of this involucre, and can lengthen up to an inch. The flowers are purplish red.

The specific name 'calcitrapa' refers to the twisted metal spikes formerly strewn across the road to impede the advance of horsemen. They were designed so that, whichever way they landed, one spike or another pointed upwards ready to damage the hoofs of the cavalry.

Despite its armament of prickles – which extend to the leaves – the Red Star-Thistle appears to be on the decline, and the *Red Data Book* of 1983 noted that since 1960 it had been reported in no more than fifteen of the 10 km squares into which the country is divided. Clapham, Tutin and Warburg's *Flora of the British Isles* has it appearing in seventeen vice-counties (out of one hundred and twelve) and places it in 'S. England from Cornwall to Northampton and Kent, South Wales and East Anglia'. It is still to be found in The Lines above Chatham in Kent where it has been known for nearly a century and a half.

It has also been recorded in Glamorgan though as a casual (which makes it a difficult plant to protect no matter where it appears). The Sussex Plant Atlas claims that the home county is now its headquarters. Even here, though, it has become scarcer than formerly. The earlier *Flora of Sussex* (1937) listed it not only along the coast from Chichester Harbour to Eastbourne, but inland for instance at Burpham to the north of Arundel. Today, in the *Sussex Plant Atlas*, it is said to grow on downs and pastures near the sea between Lancing and Beachy Head – a much smaller spread. This is generally regarded as an introduced species. Would that it could be introduced more often.

VIf. *Centaurea solstitialis* St Barnaby's Thistle

* 7th August West Sussex

According to its Latin name, the plant should be in flower at the time of the Solstice, that is about 21st June, when the sun seems to stand almost still above us on its midsummer station before returning south towards the equator. St Barnaby's Day is celebrated on 11th June but when our modern calendar was introduced on 2nd September 1752, the date was advanced by eleven days so that from then on 11th June would have occurred eleven days earlier than the season warranted, and St Barnaby's Thistle would probably still have been in bud on the saint's day. However it should have been in flower by Midsummer's Day, yet the plant which I noticed in West Sussex was only just coming into flower during the first week of August.

This is a time of year when many common tall yellow thistle-like plants are in flower – Sow-Thistles, Hawkweeds, Ox-tongues etc. – and I could well have overlooked *C. solstitialis* if I had not sat down on a piece of waste ground to clean a camera lens. And there it was beside me with all its special features – the yellow flowers rather similar, though not in colour, to those of the common Knapweed, the winged cottony stem and the bracts of the involucre edged with four spines spread out like the fingers of a hand with an extra much longer spine between them.

Centaurea melitensis (presumably from Malta), the Cockspur Star-Thistle, has even more spectacular sharply toothed bracts, but is even less often seen than St Barnaby's Thistle.

C. solstitialis most often occurs near the coast in the south-east of Britain, although in Kent it is said to be a garden weed at Stansted, in the north-west corner of the county. The *Sussex Plant Atlas* mentions five stations near the coastline and occasionally it does persist in some. My own find was in a builder's yard in East Wittering at a time when houses were being put up there. In due course the builder completed his stint and the site where St Barnaby's Thistle grew is now called Mill Gardens in deference to a disused windmill standing nearby. But I still keep looking for another find whenever I see a field of purple lucerne or pink sainfoin, for it is said that when farmers import seeds of these two plants they sometimes get yellow Star-Thistle as a bonus.

VIg. *Campanula rapunculus* Rampion Bellflower

* 8th June West Sussex

The flowers – pale purplish blue with lobes cut only halfway down the tube – hang out more or less horizontally from the spire-like stem, like bells in full peal – one of the characteristics which distinguishes this plant from *Campanula rapunculoides*, Creeping Bellflower, the bells of which hang down as if at rest. This last species once planted in flower gardens frequently becomes a weed – a sin of which Rampion Bellflower is guiltless.

Once it, too, was cultivated in gardens – but in kitchen gardens because its roots provided a welcome addition to winter salads. Now it is seldom seen, but has managed to establish itself on hedge banks in Hampshire, Sussex, Surrey, Essex and Berkshire. The Surrey record dates from 1762; the Sussex plants, of which the photograph shows one, date from 1805.

VIh. *Himantoglossum hircinum* Lizard Orchid

* 23rd June East Sussex

Such a rare and striking orchid as this is all too often picked and taken to the local museum for identification before it can be enjoyed by admirers. Three foot high, purplish green, festooned like a maypole, it often appears unexpectedly, and alone, and stays for a year in the same place before vanishing for good.

The leaves appear in the season before this plant flowers; they last through the winter, but begin to wither as the flower stalk rises, and are often dark brown or black before the first blooms appear. Each flower has a helmet of green sepals marked inside with purple lines and dots which occasionally show through. Within the helmet are two insignificant, almost invisible, whitish petals. As the flower starts to open, the lip appears, at first coiled up like a watch-spring. As it begins to uncoil we see that it is long – up to 5 cms (2 ins) and narrow – ⅛ in or less – forked at the base, with two short arms, one on each side, pointing downwards. When the flower is fully out, the lip is extended into a loose spiral winding downwards from the flower in a counter-clockwise direction. There are often sixty or more flowers to a spike and a correspondingly large number of seeds – 32,000 according to one expert's estimate.

The botanical name for the Lizard Orchid is somewhat multi-lingual. *Himantoglossum* is Greek for 'strap-tongue' and *hircinum* is the Latin for 'goat-like' and refers to the smell of the plant. But the vernacular name Lizard was already current in John Gerard's time, for, as he explained: 'The floures which grow in this bush or tuft, be very small, in forme like unto a Lizard, because of the twisted or writhen tailes, and spotted heads'. In his *Herbal* of 1597 he showed two woodcuts, one of which he identified as the Great Goat stones, and the other as the Male Goat stones; they are both drawings of the Lizard Orchids which he said 'delight to grow in fat clay ground and are seldome in any other soil to be found'. 'But', he added, 'they are seldom or never used in physick in regard to the stinking and loathsome smell and savour they are possessed with.'

John Ray, after referring to the 'heavy and poisonous odour' of the plant told his readers where to look for it. 'From the Street named Lofield in Dartford, is a place called Fleate-Lane, and about a Bow-shot on the left-hand are several plants. Also, beyond Dartford is a place named the Brent, and on the right hand a great Highway going to a village named Grimsteed-Green, a little way on the right hand you may also find it. Observed by Mr. Rouse, an ingenious botanist, and eminent apothecary in London.'

It had been known in this area since 1641 and persisted there until about 1850 when it disappeared. More recently it has appeared in the Sandwich Bay area of Kent. Many other counties have claimed it at various times: Somerset, Berkshire, Suffolk, Cambridge, Oxfordshire, Devon, Gloucestershire, Wiltshire, Norfolk, Lincolnshire and Bedfordshire, but it seems to prefer an equitable climate near to the sea. In Sussex there is a colony, decreasing when I last heard of it, in the Camber area. There, the flowers are paler and akin to the Lesser Lizard Flower or Goatstones with a white flower described by Ray as *Orchis barbata foetida minor flore albo*. The helmets of the flowers at Camber have only the faintest tinge of green (as the photograph shows).

I might add that, although the Lizard Orchid is a distinctive, one might say, unique plant when seen alone or on the dunes, it is deceptively difficult to pick out when growing on grassland.

— VII —

The Fugitives

* (See facing pp. 162 and 163 for illustrations)

VIIa. *Isatis tinctoria* Woad

* 21st May Surrey

Nearly 120 cm (4 ft) high, with brilliant yellow flowers disposed in a massive corymb, Woad stands apart from other less well-endowed members of the four-petalled Cabbage family. A single stock of this plant can give rise to several main stems, which in turn, are much branched above, so that the general effect is that of a group of several flat-topped yellow-leaved miniature trees. The leaves are bluish-grey, stalked and wavy-edged at the base of the plant and arrow-shaped and clasp round the upper parts of the stem. The achenes are distinctive too, hanging down on their stalks like black tear-drop pendants.

The blue pigment for which Woad is famous is derived from the first-year leaves of the plant. The practice among dyers was to part-dry the leaves, after which they were crushed into a paste and allowed to ferment in the air. Later, when dyers turned to Indigo, Woad fell into disuse and was no longer cultivated.

It was Caesar in his fifth book of *The Gallic Wars* who drew attention to the fact that the ancient Britons of his day coloured their skins blue, a statement to which Pliny gave currency. Some 1,600 years later, Gerard was content to attribute the custom to the Bretons and wrote: 'In France, they call it *Glastum* which is like unto Plantaine, wherewith the Brittish wives and their daughters are coloured all over, and go native in some kind of sacrifices.' *Glastum* was indeed the botanical name for the plant until Linnaeus considered that *Isatis*, the word used for the plant by the Greeks, served better.

Although Woad was used in England more than a thousand years ago, it is assumed to be an introduced plant, and has succeeded in becoming naturalised only in two widely separated sites, both of them on cliffs. The better known of the two is a cliff above the Severn near Tewkesbury in

Gloucestershire, where, however, there is some danger from erosion. The other site is on an almost vertical chalk cliff in the Guildford area of Surrey where it has been known for nearly 200 years. There access to the foot of the cliff is protected by privately owned houses, and one doubts whether anyone descending from the top would ever be able to climb up again. The only drawback to this safe site is that the splendid plants of Woad, instead of standing upright, droop over in a somewhat distorted fashion.

Alternatively it is possible to witness a profusion of cultivated plants of Woad at the Butser Hill Farm attached to the Queen Elizabeth II Country Park, near Horndean, Hampshire. July is probably the best month for a visit.

VIIb. *Saxifraga nivalis* Alpine Saxifrage

* 6th July · Ben Lawers

The last dozen plants in our list have sought refuge in areas where there are fewer walkers, fewer skiers, fewer 4-wheel drive vehicles, fewer hang-gliders and fewer ploughs. But even the remotest ben may one day be the site of a reservoir, a hydro-electric dam, a hill-road, a TV aerial, a navigational aid or a car park. Intensive grazing can, even on the least frequented fell, destroy the under-shrubs which support many different forms of life and replace them by grasses which do not. Moor burning, unless rotational, changes the character of the uplands, and conifer plantations destroy not only moorland plants but birds such as grouse, ptarmigan, golden plover and the golden eagle which cannot survive without extensive wide-ranging feeding grounds.

We deal first with a species that haunts isolated and lonely mountain rocks. It is something of a surprise to meet on a remote mountain in the highlands a plant which one feels could be more at home on a suburban rockery. For the Alpine Saxifrage on first sighting is a little like that denizen of the grotto, namely London Pride.

The leaves are in the form of a rosette, green above and purple beneath. Each is stalked and rounded like the end of a canoe-paddle or a table-tennis bat but with a coarsely cut saw-toothed edge and gland- bearing hairs near the margins. The Alpine Saxifrage is able to survive on bare rock and in this environment the plant can spread its petals over the rock without fear of meeting competition. The single, leafless, upright stalk is also covered with glandular hairs and is often purplish. The flowers are crowded into a 'golf ball' at the top of the stem. The upper surfaces of the petals are greenish-white, the lower surface reddish or purplish. The sepals too are often purple, and the anthers a muted orange.

This is a plant which is rightly described as very local or rare, but, because it has been found in sixteen vice-counties it is considered officially not to be in danger. The localities where it grows are, however, sited on remote and inhospitable mountains and one wonders how often – or seldom – the records are checked. Growing by itself on the bare rock it is easily observed and no doubt often collected.

Thomas Johnson, whose expedition in Kent is mentioned elsewhere in this book, was the first to record the Alpine Saxifrage during his visit to Snowdonia in 1639. We are not certain where exactly the discovery was made, but the evidence seems to point to the region of the Devil's Kitchen. On Snowdon itself Johnson would have had to look somewhere above Cwm Glas. Alpine Saxifrage has also been found in Dumfries, Dumbarton, Morvern, Stirling, Argyll and Skye. This would suggest that it prefers a western station where there is plenty of rain. Its site on the Ben Lawers range, which lies on Tayside further to the east, is near the summit on damp and shaded rocks reminiscent of a higgledy-piggledy Stonehenge. One might suppose that the chances of producing fertilised seed in such an environment must be small but we are assured that the plants are visited by flies. Outside Britain the Alpine Saxifrage occurs in Arctic Europe and Asia and Eastern North America to which it apparently migrated via Greenland where it is still to be found. One wonders whether Arctic Saxifrage might not have been a more suitable name for this fugitive from civilisation.

VIIc. *Allium sphaerocephalon* Round-headed Leek

* 9th July Avon

At a distance the Round-headed Leek looks like a giant burr at the tip of a long, thin stem. But, closer to, we see that the head consists of a profusion of individual bell-shaped flowers which gleam like multi-cut rubies. This must be a plant at risk for it exists only on one mainland site in Britain – the famous Avon Gorge spanned by Brunel's spectacular Suspension Bridge. On the east side of the bridge lies Clifton, and on the opposite (west) bank we have Leigh Woods – the site for the beautiful and not too common Wood Vetch with its exuberant habit and purple- striped blooms.

Most of the rarities are on the Clifton side in an area stretching from

Cook's Folly Wood, south past Sea Walls, Black Rock Quarry, The Gully (Walcombe Slade) to St Vincent's Spring and finally to St Vincent's Rocks. There, almost directly beneath the bridge, the Round-headed Leek is safe. The slope there is almost sheer and the fall from the top nearly 250 feet. Loose boulders and stones abound and a great wire fence has been installed at the foot of the cliff to discourage would-be alpinists and to stop loose rocks from encumbering the Portway which runs alongside the River Avon.

Other strange and rare Leeks growing nearby have been introduced – *Allium carinatum*, the Keeled Leek, *Allium siculum*, Honey Garlic, and *Allium roseum*, Rosy Garlic, but the Round-headed Leek is acknowledged as a native plant.

It has appeared elsewhere in the Avon Gorge, but on sites that are more easily accessible and therefore less permanent. It has also been recorded at St Aubin's Bay, Jersey.

The first ten days or two weeks of July are the best time for gazing at the Round-headed Leek, and this more or less coincides with that other attractive rarity *Veronica spicata*, the Spiked Speedwell, described elsewhere in these pages. However these two species by no means exhaust the rarities that are to be found at various times on the Gorge. One of the earliest of these is *Arabis scabra*, sometimes called *A. stricta*, Bristol Rock-cress. This flowers on the side of the Gorge as early as March. The cream-coloured flowers are relatively large for the size of the plant, and the stems, upright and usually unbranched, proceed from neat rosettes of wavy-edged leaves making the plant easily distinguishable from the many other 4-petalled cresses. *Hornungia petraea*, Rock Hutchinsia, another early-flowering rarity and *Trinia glauca*, Honewort, a later one, are to be found in almost the same spot on the same slope. But other more familiar flowers add to the charms of the Gorge – Wallflowers for instance, and naturalised red Snapdragons and Red Valerian. It seems a pity that so much of the area has been allowed to become overgrown with intrusive and for the most part unattractive shrubs.

VIId. *Astragalus alpinus* Alpine Milk-vetch

* 7th July Pitlochry

There are probably no more than four colonies of this species, the classic site being on Glen Doll which leads north-west from Glen Clova and there are other sites at Bettyhill in Caithness, in the Cairngorms and on a small mountain in the neighbourhood of Pitlochry which, however, is accessible to sheep as well as to pilgrims and suffers accordingly. One is lucky to get there ahead of the flock.

As vetches go, the plant is moderately attractive, the rows of eight to twelve pairs of leaflets being clothed with downy white hairs which give them a frosted appearance. The flowers are faintly and indeterminately coloured – whitish with a hint of bluish-purple at the tips of the petals, and one is reassured by the suspicion that gardeners are unlikely to covet it for the rockery. The caterpillars of one rare day-flying moth, the Mountain Burnet, are said however to make a meal of it.

VIIe. *Erigeron borealis* Northern Fleabane

This attractive 'daisy' with purple ray florets and yellow disc florets in the centre is found only on mountain rock ledges in Perthshire, Angus and Aberdeenshire. Even on Ben Lawers, where one expects to see so many highly prized 'Alpines' it is described as rare and local. I would probably not have noticed it there myself if the companion I was with had not reminded me that, when climbing, it is always worth turning and looking backwards and downwards as well as forward and upward, because behind and below you will see the ledge plants that were probably concealed from view when you were beneath them.

James Dickson, the Covent Garden Nurseryman who paid a visit to Ben Lawers in 1789 (not, one hopes to enrich his own stock), seems to have been the first to discover this plant.

One wonders, however, how it ever acquired its former name of Alpine Fleabane since, outside Britain it occurs in Scandinavia, Iceland and Greenland but not in the Alps.

VIIf. *Diapensia lapponica* Diapensia

It may encourage amateur botanists to know that within living memory three species new to Britain have been discovered – the most attractive of them being *Diapensia lapponica* which was found on 5th July 1951 by Mr C.F. Tebbutt (who was interested primarily in birds rather than plants).

It is a cushion-forming undershrub with dark olive-green closely packed, blunt leathery strap-like leaves and petals up to 1 cm (⅜ in). Mr Tebbutt was exceedingly lucky to see it still in flower in July because it normally blooms in May or June.

The plant grows on a single hill in the neighbourhood of Glenfinnan near Fort William on a rocky crest at a height of about 2,500 feet. The main dangers to Diapensia are the local deer and the visiting collectors.

The other two 'new' plants were *Koenigia islandica*, a minute (5 cms or 2 ins – or less) member of the Dock-Knotweed family with fleshy leaves and clustered pale green flowers, also first found in 1951. This plant grows on the Storr mountain in Skye and on the island of Mull, in the Inner Hebrides. The third plant, the *Artemsia norvegica*, Norwegian Wormwood, was discovered in Wester Ross.

VIIg. *Cicerbita alpina* Alpine Sow-Thistle

This must be one of the most conspicuous of our mountain plants, for it can grow to a height of almost 2 metres. The lowest leaves are cut in the usual manner of Sow-thistles, that is with the lobes cut pointing in reverse towards the base of the leaf, and with a pointed triangular lobe at the tip. The leaves are hairless and slightly glaucous beneath, but the stalks are clothed in red hairs. The flowers, arising from a purplish green involucre or cup of scales, are pale blue and one could, perhaps, have wished for something more striking and vivid in the way of colour to complete such a handsome structure.

The plants are among the favourite foods of sheep and deer, and survive chiefly on ledges inaccessible to them (and to all but the most intrepid botanists). Some five colonies are known, two of them on Lochnagar and Caenlochan, being accessible more or less from Braemar. But the Alpine Sow-thistle depends on a copious supply of base rich material fed to it by mountain flushes, and, where supplies become exhausted for any reason, the plants too lose their strength.

VIIh. *Lychnis alpina* Alpine Catchfly

Some people go to look for this rarity in the region of Upper Glen Clova. They take the path above Glen Doll known as Jock's Road, and strike out from it westward up the slope of the Meikle Kilrannoch, looking for the bare serpentine rock where this neat little plant, with its short-stalked pink flowers so crowded as to form almost a head, chooses to grow.

It is a perennial with a woody stock and a tap root which bores downwards, securely anchored in a crevice between two rocks. A rosette of leaves on the ground presents a low profile to the winds and a trap for whatever sunshine is offered. This was one of the plants first discovered by that remarkable field botanist George Don. He saw it in Clova in 1795 – though a specimen was not collected till 1809.

Don, born at Menmuir in Forfar in 1764, was first apprenticed to a clockmaker, but the prospect of recording hours, minutes and seconds for the rest of his life soon palled, and his love of the open air led him to take to gardening. England, he felt, offered him the best chance of gaining experience and he took several jobs there before returning to his native Scotland. There in 1797 he set up his own nursery at Dovehillok in Forfar, and began to specialise in water plants and alpines. But he was never happier than when he could set out for the hills with a bag of oatmeal, some bread and a plaid to sleep in. In 1802 he was offered, and accepted, the post of Head Gardener of the Royal Botanic Garden in Edinburgh. But he did not take kindly to the routine there and resented the restrictions placed on his freedom to go botanising in the Highlands. After four years he resigned to return to Forfar.

Glen Clova in which Don discovered *Oxytropis campestris*, Yellow Oxytropis, described elsewhere here, was only one of the many areas researched by Don. He was also at home on Ben Lawers on which he found *Myosotis alpestris*, Alpine Forget-Me-Not, on Lochnagar which provided him with *Cicerbita alpina*, Alpine Sow-Thistle, and, on Ben Lomond, where *Cerastium alpinum*, Alpine Mouse-ear, was first discovered. Between 1804 and 1812 Don published his *Herbarium Britannicarum* which appeared in quarterly issues, each describing twenty-five plants in between 1804 and 1812. Don died in 1814 but his talents were passed in some measure to his son, also George Don (1798–1856), who became foreman at the Chelsea Physic Garden.

I should add, somewhat ruefully, that *Lychnis alpina* is also to be found halfway up Hobcarton Crag near Keswick. Ruefully because it took three forays to find it. On the first occasion I felt that I could best tackle the problem by skirting the impressive crag itself, notorious for its treacherous rocks, and climbing to the top of the surrounding grass escarpment and descending from there. This tactic worked but nowhere could I see the plant. A second attempt from the ground was no more successful. But on the third occasion I happened to fall in with two friendly botanists who knew better where to look than I did – though I did happen to spot the first plant we saw. On the whole, however, the Cumbria specimens are willowy and lax in comparison with the neat little 'candlestick' Scottish types.

VIIi. *Helianthemum canum* Hoary Rock-rose

It requires a good lens to discern the finer points which distinguish this little undershrub from the more ebullient Common Rock-Rose *Helianthemum chamaecistus*. *H. canum* is smaller, growing to 20 cm (8 ins) only – about two-thirds of the size of the commoner plant. But one has to kneel down to appreciate the dense covering of whitish hairs which cover the *under*sides of the leaves and gives the plant its vernacular name. Furthermore, to be absolutely certain, one must examine the style of the flower to see whether it has an 'S' bend in the middle. If so, all is well.

This plant favours rocky limestone grassland and is found in Glamorgan, in the Gower peninsula, Pembroke, Caernarvon, Anglesey, Cumbria on Humphrey Head and on Great Ormes Head in Gwynedd.

But the special rarity is *H. canum* sub-species *levigatum* which is even humbler and more prostrate and with leaves only about half the size of those on the normal *H. canum* plants. This grows only on Cronkley Fell, Upper Teesdale, on a peculiar form of crystalline outcrop known as sugar limestone. A visit to this plant involves a lengthy morning's walk uphill from Forest-in-Teesdale westwards towards the boundary with Cumbria. But there are other treasures to be seen in the same area such as *Dryas octopetala*, Mountain Avens, *Gentiana verna*, Spring Gentian, and *Primula farinosa*, Birds-eye Primrose.

VIIj. *Veronica spicata* Spiked Speedwell

* 8th July Avon

Few wild flowers offer as fine a massed spire of blue as this – but a distinction is made between the plants growing in dry fields in the Breckland in East Anglia, which are said to be *Veronica spicata*, sub-species *spicata*, and the much larger plants – 45 cm (18 ins) or more – to be found on limestone rocks in the western half of the country, which are known as *Veronica spicata* sub-species *hybrida*.

Of course, it is well known that, when species of plants are hybridised, the resulting progeny are often larger than the original. But I have yet to hear that the larger form of *Veronica spicata* is of mixed parentage. Indeed it does seem far more probable, as David McClintock has suggested in his entertaining *Companion to Flowers*, that the difference between the two forms is due to climate – and especially the difference in rainfall between east and west, for, as he points out, when grown together in the same garden the two strains look alike.

The Breckland type, which exists in Cambridgeshire as well as in Suffolk and Norfolk, is by far the rarer of the two forms, though both are legally protected, and there is likely to be more loss of habitat through ploughing and planting in the flatlands of the east than amid the rocks of the west.

The much finer *Veronica spicata* sub-species *hybrida* occurs in south, west and north Wales – including a site on the Great Orme, and, in Cumbria, on Humphrey Head. But the most civilised site is on the Clifton side of the Avon Gorge. Sometimes the plant grows near the cliff edge at a point where the footpath teeters almost on the brink of the gorge. But if this fails, streaks of blue can be seen, clearly through field-glasses halfway down the cliff, both from the cliff top and from the road beneath. Early July is a good date to look.

VIIk. *Polemonium caeruleum* Jacob's Ladder

* 4th June Yorkshire

This fine plant with its blue-mauve chalice-blooms has been detected as flowering in Britain in late glacial times, that is about 10,000 years ago. Not surprisingly as an ex-glacial species it generally chooses a northerly slope in a northern county on which to grow. Derbyshire is one of its favourite counties, where sites include Lathkill Dale, near Bakewell, Hipley Hill, Manifold Valley, Chrome Hill and Winnats Pass where it is to be seen in company with Meadow Cranesbill, Wild Angelica and Water Avens amid the tall grass of the damper limestone meadows.

It is also found in Staffordshire and in Yorkshire where the most publicised site is around Malham Cove, above the sources of the

VIIa.

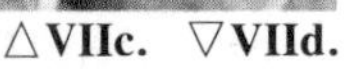

△VIIc. ▽VIId.

VIIb.

VIIa. *Isatis tinctoria* (ht up to 140 cm)
VIIb. *Saxifraga nivalis* (ht 15 cm)
VIIc. *Allium sphaerocephalon* (ht 40 cm)
VIId. *Astragalus alpinus* (ht 25 cm)

△**VIIj.** ▽**VIIl.**

VIIk.

VIIj. *Veronica spicata* (ht 30 cm)
VIIk. *Polemonium caeruleum* (ht 60 cm)
VIIl. *Pseudorchis albida* (ht 20 cm)

River Aire. The plant facing p. 163, however, was seen in Wharfdale, Yorkshire, where a small colony grows on the scree of a shady wood.

VIIl. *Pseudorchis albida* Small White Orchid

* 9th July Grampian

This small cream-coloured scented orchid – a mere 15 cm (6 ins) high – is easily overlooked, and may be commoner than is generally supposed – particularly as, even where established, it may often fail to flower. It is increasingly a northern plant, having vanished from a number of southern counties. Scotland remains its stronghold.

Hilly pastures are its favourite habitat and it is also found on moors among the heather where, as often happens, the plants are widely scattered. They are not easily refound.

The classic locality is at Keltneyburn about five miles west of Aberfeldy, but I have known it fail even there for at least one season.

The plant facing p. 163 comes from the Morrone Birkwood Reserve on the hills above Braemar and was photographed in early July. Further south the flowering time would be correspondingly earlier.

VIIm. *Phyllodoce caerulea* Blue Heath

A true blue heath would be sensational, which this plant is not. The flowers, when they appear – and they are shy – are something like those of Bell Heather – admittedly purplish – only paler, and the leaves, evergreen, are deceptively like those of the Crowberry – a very common moorland plant.

Formerly, the accepted wisdom was that this plant could be found only on a mountain known as the Sow of Atholl which lies to the west of the A9 road between Blair Atholl and Dalwhinnie in the area close to what was formerly the border between Perth and Inverness (since merged into Tayside). But the Blue Heath was originally found further north, close to what has since become the winter sports centre of Aviemore. The first record there dates from 1812, and it was discovered or rediscovered in nearby Strathspey in 1821, in 1830, and again in 1835. There followed a lapse of more than a century before the late Mary McCallum Webster saw it near Beinn Bheiol in Badenoch and Strathspey, a part of the

Highland region. Today there may be as many as six different sites, a welcome sign that as long as botanical exploration continues, with a hint of where to look, plants can become rather less rare, and rather more contemporary than our fathers might once have dreamed.

USEFUL INFORMATION

— A —

Specially Protected Wild Plants

The Wildlife and Countryside Act of 1981 strengthens the earlier Conservation of Wild Creatures and Wild Plants Act of 1975 with regard to plant protection. The number of fully protected plants (Schedule 8 of the 1981 Act) has increased from 21 to 62. These may not be picked, uprooted or destroyed. It is an offence to trade or advertise for trade in wild specimens, parts or derivatives of any of these species. The only exceptions to the provisions are where a licence is issued by the appropriate authority, for educational, conservation or other scientific purposes.

Adder's-tongue Spearwort
Ranunculus ophioglossifolius
Alpine Catchfly
Lychnis alpina
Alpine Gentian
Gentiana nivalis
Alpine Sow-thistle
Cicerbita alpina
Alpine Woodsia
Woodsia alpina
Blue Heath
Phyllodoce caerulea
Brown Galingale
Cyperus fuscus
Cheddar Pink
Dianthus gratianopolitanus
Childling Pink
Petrorhagia nanteuilii
Clove-scented Broomrape
Orobanche caryophyllacea
Diapensia
Diapensia lapponica
Dickie's Bladder-fern
Cystopteris dickieana
Downy Woundwort
Stachys germanica
Drooping Saxifrage
Saxifraga cernua
Early Spider-orchid
Ophrys sphegodes
Fen Orchid
Liparis loeselii
Fen Violet
Viola persicifolia
Field Cow-wheat
Melampyrum arvense
Field Eryngo
Eryngium campestre
Field Wormwood
Artemisia campestris
Ghost Orchid
Epipogium aphyllum
Greater Yellow-rattle
Rhinanthus serotinus
Jersey Cudweed
Gnaphalium luteoalbum
Killarney Fern
Trichomanes speciosum
Lady's-slipper
Cypripedium calceolus
Late Spider-orchid
Ophrys holoserica (fuciflora)
Least Lettuce
Lactuca saligna
Limestone Woundwort
Stachys alpina
Lizard Orchid
Himantoglossum hircinum
Military Orchid
Orchis militaris
Monkey Orchid
Orchis simia
Norwegian Sandwort
Arenaria norvegica

Oblong Woodsia
Woodsia ilvensis
Oxtongue Broomrape
Orobanche loricata
Perennial Knawel
Scleranthus perennis
Plymouth Pear
Pyrus cordata
Purple Spurge
Euphorbia peplis
Red Helleborine
Cephalanthera rubra
Ribbon-leaved Water-plantain
Alisma gramineum
Rock Cinquefoil
Potentilla rupestris
Rock Sea-lavender (two rare species)
Limonium paradoxum/Limonium recurvum
Rough Marsh-mallow
Althea hirsuta
Round-headed Leek
Allium sphaerocephalon
Sea Knotgrass
Polygonum maritimum
Sickle-leaved Hare's-ear
Bupleurum falcatum
Small Alison
Alyssum alyssoides
Small Hare's-ear
Bupleurum baldense
Snowdon Lily
Lloydia serotina
Spiked Speedwell
Veronica spicata
Spring Gentian
Gentiana verna
Starfruit
Damasonium alisma
Starved Wood-sedge
Carex depauperata
Teesdale Sandwort
Minuartia stricta
Thistle Broomrape
Orobanche reticulata
Triangular Club-rush
Scirpus triquetrus
Tufted Saxifrage
Saxifraga cespitosa
Water Germander
Teucrium scordium
Whorled Solomon's-seal
Polygonatum verticillatum
Wild Cotoneaster
Cotoneaster integerrimus
Wild Gladiolus
Gladiolus illyricus
Wood Calamint
Calamintha sylvatica

The Act also forbids the uprooting of ANY wild plant unless you are the owner or occupier of the land upon which it grows, or anyone acting with the owner's permission.

— B —

Addresses for those interested in Conservation

Botanical Society of the British Isles,
c/o The Department of Botany,
British Museum (Natural History),
Cromwell Road,
London SW7 5BD

The Countryside Commission,
John Dower House,
Crescent Place,
Cheltenham,
Gloucestershire GL50 3RA

The International Union for
Conservation of Nature and
Natural Resources,
Secretariat,
Avenue du Mont Blanc,
CH 1196 Gland,
Switzerland

The National Trust for Places of
Historic Interest or Natural
Beauty,
42 Queen Anne's Gate,
London SW1H 9AS

The National Trust for Scotland,
5 Charlotte Square,
Edinburgh EH2 4DU

The Nature Conservancy Council,
Northminster House,
Peterborough PE1 1UA

The Royal Society for Nature
Conservation,
The Green,
Nettleham,
Lincoln LN2 2NR

The Wild Flower Society,
68 Outwoods Road,
Loughborough,
Leicestershire LE11 3LY

The World Wildlife Fund,
11 Ockford Road,
Godalming,
Surrey GU7 1QU

— C —

The Wild Flower Diary

Run by the Wild Flower Society

Members of the Society can, if they wish, record their finds in the *Wild Flower Diary*. It lists the names of 1,000 of the commoner plants in the British Isles likely to be recognised by amateurs: there are blank pages on which additional species may be entered. Every plant entered in the diary must be wild, i.e. unplanted and uncultivated, and must have been seen growing by the competitor on the date shown in the diary (except disabled members who may enter any flower brought or sent to them as well as those they find themselves), and must appear in the List of British Vascular Plants by J.E. Dandy obtainable from the British Natural History Museum. Except in the cases of members with special qualifications (who can be presumed to be capable of identifying plants even when they are not in flower), plants recorded in the Diary must normally have one flower or floret fully developed on the date when seen growing and shown in the Diary.

The exception to this general rule are clubmosses, quillworts, horsetails, ferns, hornworts, elms, poplars, willows, Chinese Mugwort, and a relative of the Ragwort known because of its twining habit as German Ivy. These are plants of which the flowers are either inaccessible, invisible or irrelevant. Rushes, sedges and grasses may be counted in fruit.

On first joining the society members are allocated on a geographical basis to a branch, which may be either an 'ordinary' branch, a junior branch (for those under 18) or a group branch (to include, for example, schoolchildren, Scouts, Girl Guides and families with a child under 18). 'Groups' are of between 3 and 10: members who pay a single subscription.

The member of an ordinary branch sending in the highest number of entries in that branch moves up to join a Winners' Branch, provided that he has recorded at least 600 plants during that year. If a member has sent in good diaries for several years but has failed to gain first place in a branch, the General Secretary on recommendation by the Branch Secretary may allow an exceptional promotion for that member to a Winners' Branch. There are certain branches whose members add only new finds to their diaries each year after the first. These include Branch X: Arcadia, open to those with some sort of disability; the so-called Lotus Eaters who have sent a diary for at least two years without achieving promotion to a Winners' Branch; the Valhalla Branch whose members have been in a Winners' Branch and while there have twice recorded at least 700 plants in a year. Members of Valhalla may enter plants that are not in flower provided that there are other means of identification. They retire from the branch on recording 2,000 plants.

Juniors who are aged 12, have been members for at least two years and have recorded at least 400 plants in one year, can, if they wish join the Progressive Branch. The Dent Prize, a sum of money to be spent on a book or books on flowers is awarded from time to time to the most promising junior member. Small prizes may also be given in junior branches.

Applications to join the Society, and all payments for Diaries, should be made to the Wild Flower Society.

— D —

Maps and Map-Reading

The Ordnance Survey have produced a series Outdoor Leisure Maps. These show not only the geographical features, motorways, permitted paths, rights of way, vegetation, and landmarks, but car-parks, camping and caravan sites, mountain rescue centres, picnic sites, viewpoints, public telephone boxes and the like. But not all areas merit such treatment, and, for the others, one has to rely on the work-a-day Ordnance Survey maps. For surveying an area as a whole, use the Landranger Series. This is on a scale of 1: 50 000 which corresponds to about 1½ ins to the mile, or 2 cm to 1 km. The maps are divided into squares, measuring 1 km x 1 km which gives a fair indication of how much motoring – and walking – is likely to be involved in a botanical expedition. The Landranger series of maps cover the whole country and are distinguished by reference numbers beginning with No. 1 in the North Shetlands and finishing with No. 203 in the Isles of Scilly. Each map covers an area of roughly 40 km x 40 km.

However, the Landranger series of maps is not sufficiently detailed for 'close-up' work, and, for this, use Ordnance Survey maps on the larger scale of 1:25 000, equivalent to 4 cm to each kilometre or about 2½ ins to each mile. These maps are ruled in squares measuring 4 cm x 4 cm and show even the field boundaries and the courses of small streams.

Maps on this scale have been published in two batches known as the First Series (Provisional Edition) and the Second Series. The First Series is based on pre-1939 6-in maps revised to 1965 with subsequent major developments, road changes, etc., added with each reprinting. Each map covers an area of approximately 10 km x 10 km.

In the second series of 2½ in maps, which will eventually replace the first series, each map covers two areas, each of 10 km x 10 km, one immediately to the east of the other. Both the first and second series fit into a framework known as the National Grid.

The Grid divides Britain into squares measuring 100 km x 100 km and each of these squares is identified by two letters of the alphabet (for example TQ for the areas round London, TR for that around Dover, SH for Anglesey and so on). These alphabet squares are, in turn, divided into 100 smaller squares each corresponding to a map of 10 km x 10 km numbered from 00 to 99. Each pair of numbers shows the map's position in that alphabet square in relation to the '00' map, which is always the one at the bottom left-hand corner (south-west) of the alphabet square. When giving a map reference, the east-west position is stated first and the north-south follows after.

Thus TR10 indicates the map one place to the east of TR00, TR20, two places to the east, and so on. TR01 is the map immediately to the north of TR00, TR11 is immediately north of TR10, and TR21 is immediately to the north of TR20.

Each of these 10 km x 10 km maps are similarly divided into 100 numbered squares, each measuring 1 km x 1 km. Thus if we take the Ordnance Survey sheet for Cheddar which is ST 45, we see that the starting point on that map for measuring the distance east is 40, and that the distance to the east is marked off in kilometres from 40 to 49 inclusive, along the bottom of the map. Similarly the 5 in ST 45 indicates that the distance to the north is marked off on the side edge of the map beginning at 50 and ending at 59.

We can thus give a map reference for any landmark on that mark correct to the nearest kilometre. Also by using a set-square, we can divide each of the sides of any of the 1 km x 1 km squares into tenths, giving one hundred smaller squares and a map reference accurate to the nearest 100 metres. Thus if the spot where a flower has been found happened to lie exactly in the middle of the top right-hand kilometre square of the ST 45 map, the correct six-figure map reference would be ST 495 595, 495 being the distance east and 595 the distance north.

Still larger scale maps known as 'Six-inch' maps are available for some areas but are less frequently revised and may be of limited value. In addition to the Ordnance Survey, a number of private firms have produced large-scale maps of some areas not obtainable in the Ordnance Survey series.

SCOTTISH TERMS FOUND ON MAPS

aber river mouth
airidh shelter
ath ford
auchter upper
beag or *beg* little
bealach pass – akin to the Welsh *bwlch*
beinn, ben mountain
bideann peak
buachaille shepherd or cowman
carse broad fertile valley
clachan inn, village
col depression on a mountain or hill
corrie circular hollow on the mountain side
craig crag
drum ridge
eilean island
glas green
inver river-mouth or junction
innis, inch island
kin, ken or *caen* head
kyle (from *caol*) narrow
lag, logan, logie hollow
lairig pass
linne, linnhe pool or channel
lochan small loch
mains farmhouse
mad rounded hill
mor, more big
riach grey
ros cape or promontory
rubha, rhu or *ru* ditto

shieling shelter or hut
strath cultivated valley
struan stream
stac small mountain
tualach, tully,
tulloch hill
voe small bay or inlet
wick, vik, uig bay or cove

WELSH TERMS

aber the juncture of one stream or river with another, or with the sea.
afon river
bach little or small
bedd grave
bettws or *betws* chapel
blaen end
bwlch pass or gorge
bryn small hill
cader seat or chair
caer fort
capel chapel
carned heap of stones
cerrig stones
cefn ridge
clogwyn cliff or precipice
coch red
coed trees
craig crag
crib ridge
croes cross
cwm valley on the side of a hill
dinas fortification
dol meadow
du black
glas green
gogof cave
gwyn white
llan enclosure esp. near a church
llyn lake
maen rock
moel bare rounded mountain, headland
mynydd mountain
nant stream, glen
pen head
plas large house
rhos moor
rhyd ford
tal front
twll hollow

— E —

Taking Photographs of Wild Flowers

There is much to be said for taking photographs of flowers, even poor ones, provided that there is no risk to the flowers. Photographs, of course, cannot tell the whole story. Frequently they fail to show the fine distinctions between one species and another, in the way that line drawings can do. On the other hand, drawings or paintings in books appear against an artificial flat white or grey background. Often – horrid thought – they show the plant, roots and all – details which the law-abiding botanist will have no chance of verifying for himself. And many of the illustrations will have been concocted from a dried specimen in a herbarium, with the result that they present a rigid appearance not in accordance with their natural habit and sometimes a false coloration. Moreover in order to fit the page, stalks have to be interrupted; there are magnified corollas and berries dangle on top of the plant like the old-fashioned pawnbroker's sign; large leaves are superimposed in outline across the stem.

Photographers do not have to – indeed cannot – rely on such artifices. But they have first to make up their mind. Should they go for slides to be shown on a projector or for colour prints to be put in an album?

There is no doubt that a livelier picture is obtained with a slide since the light shines through it. You do not get this brilliance with prints because the colour on the negative has to be transferred first on to paper which then reflects the image on to the eye of the beholder. In other words, the print is removed one degree further from the original object that the camera saw. It is possible, of course, to make the prints from slides – at a cost. And those prints may not be any better than those which you would have got originally from negatives.

Also, when it comes to viewing slides, a certain amount of preparation is required. The slides have to be selected, the projector and cable installed and trained on the silver screen, and so on. In contrast, the colour-negative man can so easily flip over the pages of colour prints in his self-stick album and select those prints which would make good Christmas cards or coasters for the dinner or drinks table.

Having decided on the end-product, the photographer next needs to choose his camera. Almost any camera will allow the botanist to bring back something from an expedition. Sometimes the most telling pictures are of the environment itself – of the winding country lane fringed with hedge parsley – or the gorse on a forbidding crag, a loch mirroring the blue sky or the hills around Beddgelert crimson with rhododendrons. All these are the better for being taken through the wide-angle lens which is normally a feature of small cameras that don't cost the earth. But to get on terms of intimacy with any plant, a camera of which the lenses can be changed is unfortunately a must. For close-ups of individual blooms or sprays there is a so-called macro-lens which allows the camera to approach within about 23 cm (9 ins) of the subject. Then, if you are having ordinary en-prints made from standard 35 mm film, your colour prints will be more or less life-size. The macro-lens will probably have a focal length of 50 mm, giving approximately the angle of vision of the human eye. But then, if you want to include a plant which is, say, 2 feet high, you will do better with a wide-angle lens of say 28 mm or 24 mm. So you take off the macro – no trouble really – and fit on the wide angle instead. Later you may espy a plant that you can't reach either because it is in the middle of a lake, or on an overhanging crag. (Some of the best plants survive in this way.) In that case, a telephoto lens of 300 mm or 500 mm is called for.

For this kind of work the conventional telephoto lens weighs too much to be steadied without a tripod – making it useless for most expeditions. Instead, the itinerant photographer turns to what is known as a mirror lens which does the same job and weighs only about a quarter as much. The only disadvantage is that with mirror lenses you cannot alter the aperture. The aperture remains fixed usually either at f.5·6 or f.8. Therefore, when using the mirror lens, you either set your own camera at f.5·6 on manual, or, if your camera has aperture priority, on automatic programme at that stop. In the latter case, the camera will then automatically give the length of exposure appropriate to that stop.

By using high performance film it will usually be possible to reduce the risk of camera-shake by shortening the length of the exposure. Risk of shutter-shake can also be reduced by using the self-timer so that there is no pressure by hand on the shutter button at the time the picture is taken.

When using long-distance lenses, camera-shake is blown up just as the image itself is magnified. But a handy stone wall, or a small bean bag on the ground, or a small table-top tripod can be used to steady the camera if the light calls for an exposure of longer than say $\frac{1}{300}$ of a second using a 300 mm lens, or $\frac{1}{500}$ of a second with a 500 mm lens. Much depends on the sensitivity of the film being used. Film of a sensitivity as high as ASA 1600 is now obtainable and will usually do the trick. On the other hand, be warned! You may find when photographing on a mountain or some other wide-open space, that the light is too strong, and that the only way of cutting it down sufficiently is to screw a neutral light filter to the front of the lens.

Correct focusing is one of the difficulties that the average amateur encounters. In a camera which focuses through the lens of the camera, the picture in the view-finder appears with all the light that the camera is capable of attracting with its largest stop. Nevertheless it is often hard to see whether the image is going to be sharp. There are various types of focusing screens to make the job easier. Two circles, one within the other, occupy the centre of the focusing screen. Within the inner circle is what is known as a split image. That is the object on which you intend to focus, for example the stem of a flower or a leaf, split into two halves which come into alignment with each other only when the lens is in correct focus. Outside the split image area is an outer ring consisting of micro-prisms which shimmer like sequins or sparkling water until the focus is correctly adjusted. For additional certainty it is possible to buy a magnifier which fits over the window of the view-finder and enlarges the detail of the split image, making it easier to see when the two halves of the split image are an exact fit.

These aids will already be familiar to hardened photographers, and this is not the place to discuss the normal hazards of photography which are really the province of books and magazines devoted exclusively to that subject.

However, there are one or two problems connected especially with flower photography which can usefully be discussed. The first is that of colour. Many brightly coloured saucer-shaped flowers – in particular the heavenly blue but seldom-seen *Myosotis alpestris*, Alpine Forget-Me-Not, reflect the light upwards, so that instead of heavenly blue, the camera is given back whatever white light is coming out of the sky above. The same problem occurs with yellow flowers, including even those such as *Inula helenium*, Elecampane, which are more like a dish than a mop. Some relief can be obtained by photographing the flower sideways on or against the light. It may then be possible to arrange for the light to reach the camera through two layers of petal, thus doubling the strength of the colour. Shooting from the side avoids direct reflections and shooting against the light often ensures that the bloom itself will be slightly underexposed and thus deeper coloured.

Another difficulty occurs only too frequently when a plant has to be photographed against a dark shadowy background. Then, if the camera is on automatic, the exposure given by the camera will be based on the dark background and the flower itself will appear over-exposed and colourless. A number of ways have been devised for overcoming this problem. Some cameras have centre-weighted metering, that is the exposure reading takes more account of the amount of light coming from objects in the centre of the picture, than from objects at the extremities. Other cameras can take light readings exclusively from the centre of the picture or even from a given spot. Such cameras often have a 'memory-hold' facility. This allows a reading taken close up to the plant to be held by pressing a memory button, in which case the shutter and aperture will not change even when the position of the camera is altered as would happen normally with an automatic camera.

There are also other ways of 'taming' the automatic camera into doing what the operator wants. Thus if a flower is being taken against a dark background and you therefore wish for a shorter exposure than the camera would automatically give, then you can shorten the exposure by pretending that you have a faster film in the camera than is actually the case. So if the camera is set for film with a sensitivity of ASA 400, you can adjust the reading on the camera to ASA 800. Naturally you go into reverse when photographing a plant against a white sky, when you will need a longer exposure than the automatic camera would give you.

Wind is a permanent enemy of good photography. On some cameras – those that have 'shutter priority' you can set the shutter at the fast speed you need on such occasions – say 1/1000th of a second, and the camera, if on automatic, will find the right aperture – even if the sun comes out suddenly from behind the cloud. Without shutter priority you might get a slower shutter speed – which would be unfortunate – as soon as the sun appeared. Another possibility is to freeze the movement of the plant in the wind by using flash-light. But this is really second-best; for photographs taken by flashlight need careful positioning to avoid the plant looking stark against a background of utter darkness.

Rain is an even more serious affair. There are, it is true, underwater cameras, and rain-proof cameras, but none of them is suitable for close-up through-the-lens focusing of the kind we have been considering. It is also possible to obtain plastic cases within which the camera can be controlled by the operator from outside using primitive gloves. One other possible arrangement is a large loose plastic bag to cover the camera and the head and shoulders of the operator. The bag has a small hole cut in the front through which the lens of the camera protrudes. The fabric of the bag is secured round the lens with rubber bands.

This would protect the camera perhaps for one vital shot or two but would be unserviceable on an expedition. In practice one keeps the camera as far as possible in its waterproof padded container, takes it out as little as possible, shields it as far as possible with one's waterproof jacket, wipes it dry as far as possible with a cloth which one tries to keep dry – and hopes for the best. A friend with a translucent umbrella would, of course, be a help. But on the whole I am inclined to wait until it stops raining.

A Conservation Code devised by the Botanical Society of the British Isles includes some good suggestions for photographers. The Society advises photographers to:

1. Watch your step. It is all too easy to step unwittingly on seedlings not yet in flower and therefore not easily distinguishable.
2. Treading on the soil may compact it, preventing water reaching seedlings of growing plants. (It is especially damaging in marshy habitats).
3. When visiting a rare plant, avoid leaving any evidence which might betray its existence – such as taking a direct path to it, or trampling down the vegetation around it. Cover your traces by restoring the vegetation to its original posture.

   'Gardening', i.e. getting rid of vegetation which obscures the camera's view can also put rarities at risk. Live grass etc. should be tied together lightly rather than cut away and released again as soon as the picture has been taken.

4. If possible stand while taking a photograph rather than kneel, and kneel rather than lie. Use a long-distance lens if there are likely to be seedlings around.
5. Avoid telling the world – particularly if you have found a rare plant yourself. Let the local Conservation Trust know so that they can see it is protected.
6. Take care that your photograph does not contain give-away clues as to the whereabouts of rarities – particularly on mountain ranges such as Snowdon where well-known peaks – not to speak of the Snowdon railway itself – are easily identifiable. Power-line pylons and lochs are another give-away, particularly in the Highlands.

SOME SELECTED DATES FOR A PHOTOGRAPHER'S DIARY*

Month	Date	Latin name	Common name
March	16th	*Helleborus viridis*	Green Hellebore
	21st	*Draba aizoides*	Yellow Whitlowgrass
April	9th	*Pulsatilla vulgaris*	Pasque Flower
	10th	*Romulea columnae*	Sand Crocus
	12th	*Gagea lutea*	Yellow Star-of-Bethlehem
	18th	*Saxifraga oppositifolia*	Purple Saxifrage
	19th	*Buxus sempervirens*	Box
	27th	*Myosurus minimus*	Mousetail
	29th	*Tulipa sylvestris*	Wild Tulip
May	2nd	*Fritillaria meleagris*	Fritillary
	9th	*Ophrys sphegodes*	Early Spider Orchid
	14th	*Orchis purpurea*	Lady Orchid
	19th	*Gentiana verna*	Spring Gentian
	21st	*Lychnis viscaria*	Sticky Catchfly
		Scandix pecten veneris	Shepherd's Needle (for seed)
		Matthiola incana	Hoary Stock
	24th	*Lonicera xylosteum*	Fly Honeysuckle
	26th	*Trifolium stellatum*	Starry Clover
	29th	*Orchis ustulata*	Burnt Orchid
June	3rd	*Dryas octopetala*	Mountain Avens
		Lloydia serotina	Snowdon Lily
	4th	*Polemonium caeruleum*	Jacob's Ladder
		Cypripedium calceolus	Lady's Slipper Orchid
	6th	*Helianthemum apenninum*	White Rockrose

*These are average dates which do not necessarily correspond to those mentioned in the text.

		Oenothera stricta	Scented Evening Primrose
	7th	*Isatis tinctoria*	Woad
	8th	*Petrorhagia nanteuilii*	Childling Pink
		Aristolochia clematitis	Birthwort
		Campanula rapunculus	Rampion Bellflower
	9th	*Ophrys fuciflora*	Late Spider Orchid
	11th	*Orchis simia*	Monkey Orchid
		Melittis melissophyllum	Bastard Balm
	12th	*Phyteuma spicatum*	Spiked Rampion
	19th	*Salvia pratensis*	Meadow Clary
	23rd	*Himantoglossum hircinum*	Lizard Orchid
	25th	*Phyteuma tenerum*	Round-headed Rampion
	30th	*Bartsia alpina*	Alpine Bartsia
July	1st	*Helianthemum canum*	Hoary Rockrose
		Potentilla fruticosa	Shrubby Cinquefoil
	4th	*Silene noctiflora*	Night-flowering Catchfly
		Gentiana nivalis	Alpine Gentian
		Saxifraga nivalis	Alpine Saxifrage
		Saxifraga cernua	Drooping Saxifrage
		Erigeron borealis	Alpine Fleabane
		Myosotis alpestris	Alpine Forget-me-not
	6th	*Dianthus deltoides*	Maiden Pink
		Veronica fruticans	Rock Speedwell
		Silene otites	Spanish Catchfly
	7th	*Damasonium alisma*	Starfruit
		Astragalus alpinus	Alpine Milkvetch
	8th	*Veronica spicata*	Spiked Speedwell
	9th	*Epipactis atrorubens*	Dark Red Helleborine
		Linnaea borealis	Twinflower
		Pseudorchis albida	Small White Orchid
		Allium sphaerocephalon	Round-headed Leek
	10th	*Melampyrum arvense*	Field Cow-wheat
	11th	*Dianthus gratianopolitanus*	Cheddar Pink
		Liparis loeselii	Fen Orchid
	12th	*Gladiolus illyricus*	Wild Gladiolus
		Ludwigia palustris	Hampshire Purslane
	13th	*Centaurea calcitrapa*	Red Star-Thistle
		Oxytropis campestris	Yellow Oxytropis
	14th	*Epipactis palustris*	Marsh Helleborine
		Matthiola sinuata	Sea Stock
	15th	*Melampyrum cristatum*	Crested Cow-wheat
	16th	*Cephalanthera rubra*	Red Helleborine
		Cirsium tuberosum	Tuberous Thistle
	17th	*Hammarbya paludosa*	Bog Orchid
	18th	*Orobanche caryophyllacea*	Clove-scented Broomrape
	23rd	*Bupleurum rotundifolium*	Thorow-wax
		Agrostemma githago	Corncockle

	25th	*Teucrium chamaedrys*	Wall Germander
August	5th	*Allium babingtonii*	Babington's Leek
	6th	*Adonis annua*	Pheasant's Eye
	7th	*Centaurea solstitialis*	St Barnaby's Thistle
	20th	*Epipogium aphyllum*	Ghost Orchid
	25th	*Gentiana pneumonanthe*	Marsh Gentian
	30th	*Inula helenium*	Elecampane
September	11th	*Pulicaria vulgaris*	Small Fleabane
	13th	*Calamintha sylvatica*	Wood Calamint
		Inula crithmoides	Golden Samphire
	15th	*Mentha pulegium*	Pennyroyal
	20th	*Aster linosyris*	Goldilocks Aster

— F —

Glossary
of Commonly Used Botanical Terms

Achene	Dry, one-seeded fruit
Annual	A plant that lives for one year only
Anther	The part of the stamen containing its pollen
Axil	The angle between the upper surface of a leaf and the stem
Base-rich	Soil containing mineral salts; not acidic
Biennial	A plant with a life-cycle of two years, flowering in the second
Bogs	Based on wet acid peat and sphagnum moss
Bract	Leaf-like growth beneath a flower or flowers
Bracteole	Leaf-like growth beneath a single flower
Bulb	Underground modified stem with leaves enclosing next year's bud
Bulbil	Small bulb formed above ground
Calyx	Collective name for the *sepals*
Casual	A non-native species occurring sporadically but not persisting
Chlorophyll	The green pigment in leaves etc. which enables a plant to convert light energy into chemical energy
Corm	An erect and swollen underground stem which dies down at the year's end, giving rise to a new growth the following season
Corolla	Collective name for the petals
Corymb	A flat-topped *inflorescence*
Fens	Marshes with organic soil which may be either alkaline, neutral or slightly acid
Fertile	Producing viable pollen or seed capable of germination
Filament	The stalk supporting the anther
Flush	Wet ground over which water streams
Heath	Area dominated by heather tribe: sandy or with a light covering of peat
Inflorescence	The flowering system of a plant
Involucre	Bracts resembling a calyx
Lanceolate	Leaves at least three times as long as broad, coming to a point

Marsh	Area in which the summer water-level is at or near the surface. Based on inorganic clay or silt in contrast to fens and bogs
Meadow	Field reserved for hay-crop
Moor	Dry upland heath
Native	Not known to have been introduced by human agency
Node	A point on the stem from which one or more leaves grow
Ovule	The egg which will develop into a seed following fertilisation
Pedicel	The stalk of a single flower
Peduncle	The stalk of an inflorescence
Perennial	A plant that survives more than two years, flowering annually after the first year
Petals	The inner segments of the flower
Petiole	The stalk of a leaf
Raceme	A growing spire of stalked flowers with the youngest and the shortest stalks at the top
Ray	In daisy-type species, the flower heads often contain two different forms of flower: the central, usually tubular, flowers known as disc florets and the marginal flowers with strap-like ligules known as ray florets
Rhizome	An underground, usually thickened, and creeping stem
Sepals	The outer segments of the flower enclosing the petals, if any
Sessile	Having no stalk
Simple	Leaves without leaflets
Sinuate	Wavy
Sinus	The inlet between two lobes of a leaf
Spike	A *raceme* of *sessile* flowers
Spur	A tubular projection at the base of a petal or calyx
Stamen	The 'male' organ consisting of a *filament* and *anther*
Sterile	The opposite of *fertile*
Stigma	The area on top of the *style* which receives the pollen
Stipule	Leaf-like growth usually below the leaf-stalk
Style	The stalk connecting the ovary to the *stigma*
Tap root	A strong main root striking down vertically
Tuber	A swollen part of the stem or root not arising directly from a previously formed tuber
Umbel	The formation in which all the flower-stalks arise from the same point on the stem, umbrella-fashion
Whorl	Formation in which three or more leaves grow around the stem at the same level

— G —

Select Bibliography

Allen, D.E., *The Botanists: A History of the Botanical Society of the British Isles,* St Pauls Bibliographies, 1986

Anderson, P. and Shimwell, D., *Wild Flowers and other Plants of the Peak District,* Morland Publishing, 1981

Arnold, Rev. F.H., *Flora of Sussex,* Simpkin Marshall, Hamilton Kent, 1887

Barr, Colin, Benefield, Chris, Bunce, Bob, Ridsdale, Heather, Whittaker, Margaret, *Landscape Changes in Britain,* Institute of Terrestrial Ecology, 1986

Benefield, Chris, see Barr, Colin

Bentham, George, Hooker, Sir J.D., Rendle, A.R. *Handbook of the British Flora* (with illustrations by W.H. Fitch, W.S. Smith and others 1924), L. Reeve, 1945

Bevis, J.H., Kettell, R.E., Shepard B., *Flora of the Isle of Wight,* Isle of Wight Natural History and Archaeological Society, 1978

Bradshaw, Dr. M.E. (ed.), *The Natural History of Upper Teesdale,* Durham County Conservation Trust, 1976

Bridson, Gordon, see Stearn, W.T.

Bunce, Bob, see Barr, Colin

Campbell, Bruce, see North, F.J.

Clapham, A.R., Tutin, T.G., Warburg, E.F., *Flora of the British Isles,* Cambridge University Press, 1962

Clements, M.A., Muir, H., Cribb, P.J., 'A Preliminary Report on the Symbiotic Germination of Terrestrial Orchid Species', *Kew Bulletin,* 1986

Cribb, Dr. P.J., see Clements, M.A.

Dony, J.G., Perring, F.H., Rob, C.M., *English Names of Wild Flowers,* Botanical Society of the British Isles, 1980

Druce, G.C., *Comital Flora of the British Isles,* Arbroath, 1933

Ellis, Gwynn, *Plant Hunting in Wales,* Bulletins of the National Museum of Wales, 1972–1974

——*Flowering Plants of Wales,* National Museum of Wales, 1983

Ewen, A.H., Prime, C.T. (transl. Co ed.), *Ray's Flora of Cambridgeshire,* Wheldon & Wesley, 1975

Farrell, Lynne, see Perring, F.H.

Fitter, R.S.R., *Finding Wild Flowers,* Collins, 1971

Fitter, Richard, Fitter, Alastair, & Blamey, Marjorie, *The Wild Flowers of Britain and Northern Europe,* Collins, 1974

Gerard, John, *The Herbal* (edition of 1633), Dover Publications, facsimile reprint 1970
Gilmour, J.S.L., *Thomas Johnson's Botanical Journeys,* Hunt Botanical Library, 1972
Hall, P.C., *Sussex Plant Atlas,* Booth Museum of Natural History, Borough of Brighton, 1980
Hooker, Sir J.D., see Bentham, George
Hywel-Davies, Jeremy, & Thom, Valerie *The Macmillan Guide to Britain's Nature Reserves,* Macmillan, 1984
Ingram, Ruth, & Noltie, Henry J. *The Flora of Angus,* Dundee Museum Art Galleries, 1981
Johns, Rev, C.A., *Flowers of the Field,* Society for Promoting Christian Knowledge, 1892
Johnson, Thomas, *A Journey for the Investigation of Plants in the County of Kent,* Hunt Botanical Society, Pittsburgh, 1972
Kettell, R.E., see Bevis, J.H.
Lees, F. Arnold, *The Vegetation of Craven,* T. Buncle, 1939
Linnaeus, Carl, *Flora Anglica* (1754 and 1759), The London Ray Society, 1973
Lousley, J.E., *Wild Flowers of Chalk and Limestone,* Collins, 1969
——*Flora of Surrey,* David & Charles, 1976
McCallum Webster, Mary, *Flora of Moray, Nairn and East Inverness,* Aberdeen University Press, 1977
Martin, W. Keble, *The Concise British Flora in Colour,* Ebury Press/Michael Joseph, 1969
——*Sketches for the Flora,* Michael Joseph/Flora Martin, 1972
McClintock, David, *A Companion to Flowers,* Bell, 1966
McClintock, David, & Fitter, R.S. *The Pocket Guide to Wild Flowers,* Collins, 1956
Muir, H., see Clements, M.A.
Nature Conservancy Council, *Annual Reports*
Noltie, Henry J., see Ingram, Ruth
North, F.J., Campbell, Bruce, & Scott, Richenda *Snowdonia,* Collins, 1949
Parkinson, John, *Paradisi in Sole Paradisus Terrestris,* Methuen facsimile reprint, 1904
Perring, F.H., Farrell, Lynne, *British Red Data Book, Vascular Plants* (2nd ed.), Royal Society for Nature Conservation, 1983
Perring, F.H., with Sell, P.D., Walters, S.M., & Whitehouse, H.L.K., *A Flora of Cambridgeshire,* Cambridge University Press, 1964
Philp, Eric G., *Atlas of the Kent Flora,* Kent Field Club, 1982
Prime, C.T., see Ewen, A.H.
Proctor, Michael, & Yeo, Peter, *The Pollination of Flowers,* Collins, 1979
Raven, John, Walters, Max, *Mountain Flowers,* Collins, 1956
Ray, John, *Synopsis Methodica Stirpium Britannicarum* (1724) The London Ray Society, 1973
Rendle, A.R., see Bentham, George
Ridsdale, Heather, see Barr, Colin
Rob, C.M., see Dony, J.G.

Rose, Francis, *The Wild Flower Key,* Frederick Warne, 1981
Scott, Richenda, see North, F.J.
Sell, P.D., see Perring, F.H.
Shepard, B., *Supplement to the Flora of the Isle of Wight,* see also Bevis J.H. Isle of Wight Natural History and Archaeological Society, 1983
Shimwell, D., see Anderson, P.
Smith, A.E. (ed.), *A Nature Reserves Handbook* Royal Society for Nature Conservation, 1982
Stearn, Prof. W.T., & Bridson, Gordon *Carl Linnaeus 1707–1778* The Linnean (sic) Society of London
Summerhayes, V.S., *Wild Orchids of Britain,* Collins, 1985
Synge, Hugh, *The Biological Aspects of Rare Plant Conservation* John Wiley, 1985
Thom, Valerie, see Hywel-Davies, Jeremy
Tutin, T.G., *Umbellifers of the British Isles,* Botanical Society of the British Isles, 1980
Walters, M., see Raven, J., see also Perring, F.H.
Warburg, E.F., see Clapham, A.R.
Whitehouse, H.L.K., see Perring, F.H.
Whittaker, Margaret, see Barr, Colin
Wolley-Dod, Lt. Col. A.H., *Flora of Sussex,* Kenneth Saville, 1937
Yeo, Peter, see Proctor, Michael

Index of Plant Names

(Page numbers in *italic* indicate main references)

Index of People and Places